Control and Filter Design of Single-Phase Grid-Connected Converters

IEEE Press
445 Hoes Lane
Piscataway, NJ 08854

Control and Filter Design of Single-Phase Grid-Connected Converters

Weimin Wu
Shanghai Maritime University, Shanghai, China

Frede Blaabjerg
Aalborg University, Aalborg, Denmark

Henry Chung
City University of Hong Kong, Kowloon Tong, Hong Kong

Yuanbin He
Hangzhou Dianzi University, Zhejiang, China

Min Huang
Shanghai Maritime University, Shanghai, China

IEEE PRESS

WILEY

Published by John Wiley & Sons, Inc., Hoboken, New Jersey.
Published simultaneously in Canada.

For general information on our other products and services or for technical support, please contact our Customer Care Department within the United States at (800) 762-2974, outside the United States at (317) 572-3993 or fax (317) 572-4002.

Wiley also publishes its books in a variety of electronic formats. Some content that appears in print may not be available in electronic formats. For more information about Wiley products, visit our web site at www.wiley.com.

Library of Congress Cataloging-in-Publication Data
Names: Wu, Weimin (Professor), author. | Blaabjerg, Frede, author. | Chung, Henry Shu-hung, author. | He, Yuanbin, author. | Huang, Min (Assistant professor), author.
Title: Control and filter design of single phase grid-connected converters / Weimin Wu, Frede Blaabjerg, Henry Chung, Yuanbin He, Min Huang.
Description: Hoboken, New Jersey : Wiley-IEEE Press, [2023] | Includes bibliographical references and index.
Identifiers: LCCN 2022046741 (print) | LCCN 2022046742 (ebook) | ISBN 9781119886549 (cloth) | ISBN 9781119886570 (adobe pdf) | ISBN 9781119886587 (epub)
Subjects: LCSH: Electric inverters. | Distributed generation of electric power.
Classification: LCC TK7872.I65 W826 2023 (print) | LCC TK7872.I65 (ebook) | DDC 621.3815/322–dc23/eng/20221018
LC record available at https://lccn.loc.gov/2022046741
LC ebook record available at https://lccn.loc.gov/2022046742

Cover Design: Wiley
Cover Image: © Somsit/Shutterstock

Set in 9.5/12.5pt STIXTwoText by Straive, Pondicherry, India

Contents

Author Biography

Weimin Wu (M'16) received his BS degree in electrical engineering from Anhui University of Science & Technology Anhui, China, in 1997; his MS degree in control theory and control engineering from Shanghai University, Shanghai, China, in 2001; and his PhD degree in electrical engineering from the College of Electrical Engineering, Zhejiang University, Hangzhou, China, in 2005.

He worked as a research engineer in the Delta Power Electronic Center (DPEC), Shanghai, from July 2005 to June 2006. Since July 2006, he has been a faculty member at Shanghai Maritime University, where he is currently a full professor in the Department of Electrical Engineering. He was a visiting professor in the Center for Power Electronics Systems (CPES), Virginia Polytechnic Institute and State University, Blacksburg, USA, from September 2008 to March 2009. From November 2011 to January 2014, he was also a visiting professor in the Department of Energy Technology, Aalborg University, Denmark, working at the Center of Reliable Power Electronics (CORPE). He has coauthored over 200 papers and holds 18 patents. His areas of interest include power converters for renewable energy systems, power quality, smart grid, and energy storage technology. Prof. Wu serves as an associate editor for the *IEEE Transactions on Industrial Electronics*.

Frede Blaabjerg (S'86–M'88–SM'97–F'03) was with ABB-Scandia, Randers, Denmark, from 1987 to 1988. From 1988 to 1992, he was a Ph.D. Student with Aalborg University, Aalborg, Denmark. He got his PhD degree in electrical engineering at Aalborg University in 1995. He became an assistant professor in 1992, an associate professor in 1996, and a full professor of power electronics and drives in 1998 at AAU Energy. From 2017 he became a Villum Investigator. He is *honoris causa* at University Politehnica Timisoara (UPT), Romania, in 2017 and Tallinn Technical University (TTU), Estonia, in 2018.

His current research interests include power electronics and its applications such as in wind turbines, PV systems, reliability, harmonics, and adjustable speed drives. He has published more than 600 journal papers in the fields of power electronics and its applications. He is the coauthor of four monographs and editor of ten books in power electronics and its applications.

He has received 33 IEEE Prize Paper Awards, the IEEE PELS Distinguished Service Award in 2009, the EPE-PEMC Council Award in 2010, the IEEE William E. Newell Power Electronics Award 2014, the Villum Kann Rasmussen Research Award 2014, the Global Energy Prize in 2019, and the 2020 IEEE Edison Medal. He was the editor in chief of the IEEE *Transactions on Power Electronics* from 2006 to 2012. He was a distinguished lecturer for the IEEE Power Electronics Society from 2005 to 2007 and for the IEEE Industry Applications Society from 2010 to 2011 as well as 2017 to 2018. In 2019–2020 he served as the president of the IEEE Power Electronics Society. He was the vice president of the Danish Academy of Technical Sciences.

He was nominated in 2014–2021 by Thomson Reuters to be among the most 250 cited researchers in engineering in the world.

Henry Chung (M'95-SM'03-F'16) received his BEng and PhD degrees in electrical engineering from the Hong Kong Polytechnic University, Kowloon, Hong Kong, in 1991 and 1994, respectively. Since 1995, he has been with the City University of Hong Kong, Kowloon, where he is currently Dean of Students, a chair professor at the Department of Electrical Engineering, and the director of the Centre for Smart Energy Conversion and Utilization Research.

His current research interests include renewable energy conversion technologies, lighting technologies, energy harvesting, smart grid technologies, and computational intelligence for power electronic systems. He has edited one book, authored eight research book chapters, and published over 460 technical papers including 220 refereed journal papers in his research areas, and he holds 80 patents.

Prof. Chung is recipient of the 2021 IEEE PELS R. David Middlebrook Achievement Award, the CityU Teaching Excellence Award in 2022 and 2018, respectively, and the Outstanding Research Award in 2020. He is currently associate editor of the *IEEE Transactions on Power Electronics* and the *IEEE Journal of Emerging and Selected Topics in Power Electronics*. He was the editor in chief of the *IEEE Power Electronics Letters* from 2014–2018. He was also chair of the Technical Committee of the High-Performance and Emerging Technologies, IEEE Power Electronics Society in 2010–2014. He has received numerous industrial awards for energy saving technologies he invented.

Yuanbin He received both his BS and MS degrees in electrical engineering from The Shanghai Maritime University, in 2009 and 2011, respectively, and his PhD degree in electrical engineering from The City University of Hong Kong, Hong Kong, in 2017, where he also worked as a research assistant from April to August 2013 and a postdoctoral research fellow from February to July 2017. From July 2011 to March 2013, he worked as an associate researcher in Nanjing FSP-Powerland Technology Inc., Nanjing, China, where he has been engaged in research and development of on-board charger and photovoltaic grid-connected inverters. From February to June 2016, he was a visiting scholar at the University of Manitoba, Winnipeg, Canada. Since May 2017, he has been with the Hangzhou Dianzi University, Hangzhou, China, where he is currently an associate professor in the Department of Electrical Engineering and Automation. His current research areas include renewable energy generation system, power quality, and smart grid.

Min Huang received her BS degree in electrical engineering from Anhui University of Technology, her MS degree in electrical engineering from the Shanghai Maritime University, Shanghai, China, in 2012, and her PhD degree in electrical engineering from the Institute of Energy Technology, Aalborg University, Aalborg, Denmark, in 2015. During August–November 2014, she was a visiting scholar at the University of Alberta, Canada. Since April 2016, she has been with the Shanghai Maritime University, Shanghai, China, where she is currently an assistant professor in the Department of Electrical Engineering. Her research interests include power quality, control, and power converters for renewable energy systems.

Preface

The purpose of this book is to promote a comprehensive understanding of high-order power filter design and system stability control of grid-connected inverters for senior undergraduates, graduate students, researchers, and engineers in the electrical engineering area.

Distributed generation systems are widely used in future power systems, and grid-connected inverters based on high-order filters will be important interface devices. In the single-phase system and the three-phase four-wire three-phase system, the *LLCL* filter can significantly save more grid-side inductance than the *LCL* filter. Note that the smaller grid-side inductance not only saves costs but also faces the larger control challenges caused by the large variation of the grid equivalent impedance or reactance. This book first conducts in-depth theoretical analysis on how to reasonably design *LCL* and *LLCL* filters and gives design examples; second, it studies and analyzes how to use pure passive damping, hybrid damping, and LC trap voltage active damping methods to improve the stability of the *LCL* or *LLCL*-filtered grid-connected inverter; finally, it also analyzes how to improve the performance of grid-connected inverters by adopting the method of canceling the grid impedance in series.

This book contains 10 chapters:

Chapter 1 begins with a brief introduction to the power conversion and control stages of low-voltage grid-tied inverters and the intrinsic and extrinsic challenges of controlling grid-tied voltage source inverters with higher-order power filters, and then briefly reviews how the different responses measure.

Chapter 2 introduces the control structure and modulation techniques of single-phase grid-connected inverter.

Chapter 3 analyzes the principles and design methods of *LCL* and *LLCL* filters based on the PWM voltage spectrum of a single-phase grid-connected inverter. It can be seen that the *LLCL* filter can save 38% of the total inductance compared to the *LCL* filter.

Chapter 4 presents two different *LLCL* filter measures to suppress EMI noise with almost negligible additional hardware cost.

Chapter 5 studies how to design the reasonable passive RC damper for *LCL* or *LLCL*-filtered grid-connected inverter under stiff or weak grid.

When the equivalent grid impedance changes over a wide range, Chapter 6 introduces a pure composite passive damper to ensure the stability of system, but with extra material of inductance.

In order to ensure the stability of *LCL* or *LLCL*-filtered grid-connected inverters in the case of large grid impedance variation and low cost, the hybrid damping method is introduced in Chapter 7, and the detailed design process is given.

Also, in some cases the inverter is connected to the grid via long cables. Chapter 8 presents a robust design of an impedance-based *LLCL*-filtered grid-tied inverter to cope with wide variations in grid reactance.

Different from traditional capacitor current active damping method of a *LCL*-type grid-connected inverter, in Chapter 9, a simple damper with trap voltage feedback is investigated for a *LLCL*-filtered system, where lesser noise complications and faster dynamics can be achieved.

Chapter 10 discusses an active grid impedance cancelator using the concept of a series active filter to suppress the impact of grid impedance and to stabilize single-phase grid-connected inverter with *LCL* filter under variable grid conditions. Compared with the traditional inverter, the reshaped inverter with the grid impedance cancelator performs as a well-damped one without equivalent grid impedance.

Finally, the authors hope to further promote the widespread application of higher-order filters, especially *LLCL* filters, in the field of low-voltage renewable distributed power generation systems.

Part I

Background

1 Introduction

A paradigm shift from large power plants to small distributed generation (DG) systems located at the point of consumption is an emerging trend in the field of electricity. In recent years, DG systems have been powered by photovoltaic (PV) cells, wind turbines, wave generators, fuel cells, small hydro-powered, and gas-powered combined heat and power stations [1–5]. As reported by the Global Renewable Energy Policy Multi-Stakeholder Network REN21, solar PV cells and wind are currently the mainstream options in the power sector with an increasing number of countries generating more than 20% of their electricity. As shown in Figure 1.1, the installed renewable power capacity increased beyond 200 GW in 2019 (mostly with solar PV cells), which is the largest increase ever. During the same year, 57% of renewable power capacity additions were using solar PV cells (direct current) followed by wind power (approximately 60 GW for 30%) and hydropower (approximately 16 GW for 8%). The remaining 5% of additions were from bio-power, geothermal power, and concentrating solar thermal power. For the fifth year in a row, net additions of renewable power generation capacity clearly outpaced the net installations of fossil fuel and nuclear power capacity combined as shown in Figure 1.2. Globally, 32 countries had at least 10 GW of renewable power capacity in 2019; a decade earlier, it was only 19 countries. In most countries, producing electricity from wind and solar PV cells is cost-effective than generating electricity from new coal-fired power plants.

1.1 Architecture of DG Grid-Connected Converter

Regardless of the type of renewable energy source, a DG grid-connected converter is essential for converting the energy produced by the source to the grid [2, 3]. Considering their high degree of modularity, scalability, adaptability, maintainability,

Control and Filter Design of Single-Phase Grid-Connected Converters, First Edition. Weimin Wu, Frede Blaabjerg, Henry Chung, Yuanbin He, and Min Huang.

Published 2023 by John Wiley & Sons, Inc.

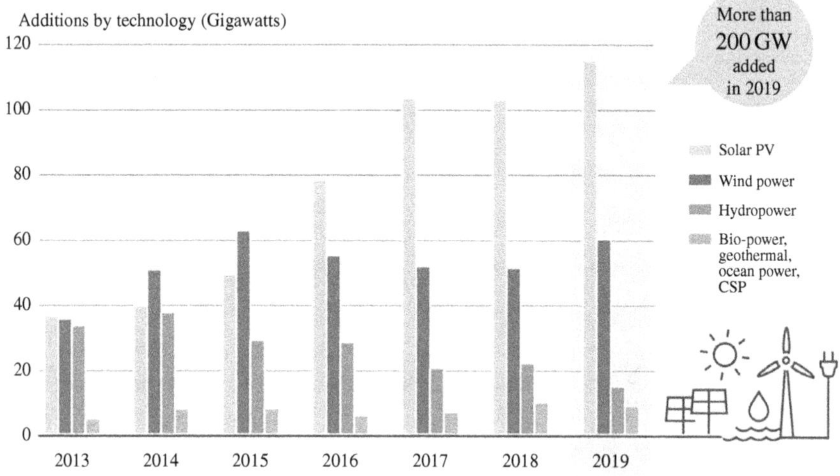

Figure 1.1 Estimated renewable share of total final energy consumption, 2018. *Source:* [1]/REN21.

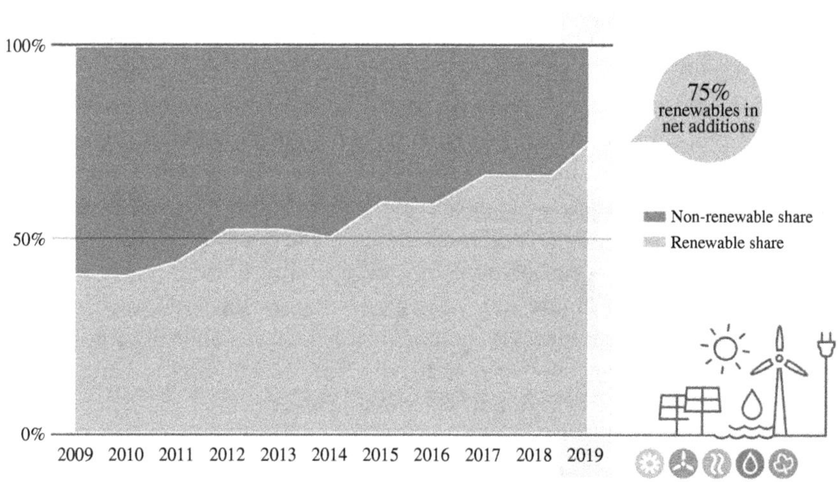

Figure 1.2 Renewable and non-renewable shares of net annual additions in power generation capacity, 2009–2019. *Source:* [1]/REN21.

and autonomic behavior, DG grid-connected converters have been widely employed in many commercial, industrial, and domestic applications.

The basic structure of a modern DG grid-connected converter is similar to that of power electronics systems. It consists of a power conversion stage (PCS) and a

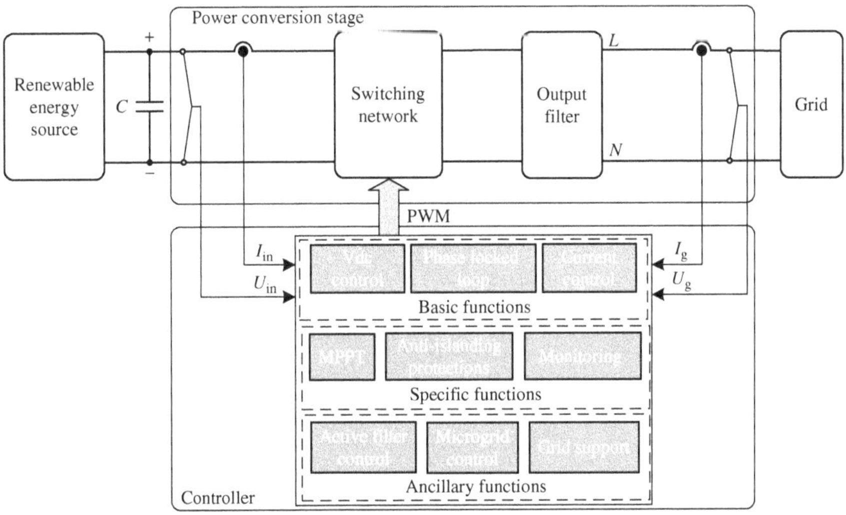

Figure 1.3 Architecture of DG grid-connected converter.

controller. Its architecture is shown in Figure 1.3. The controller includes basic functions, specific functions, and ancillary functions [4]. The basic functions are necessary to ensure normal operation of the system and specific functions are used to meet the industrial requirements. The ancillary functions are optional depending on specific applications such as microgrid applications [5]. The basic functions of the controller are typically composed of grid synchronization or phase-locked loop (PLL), DC-link voltage regulation, and output current regulation.

1.1.1 Power Conversion Stage

The PCS of DG grid-connected converter consists of a switching network and an output filter as shown in Figure 1.3. The switching network converts incoming DC power into high-frequency AC power. The output filter allows the line-frequency component to pass and attenuate, thus switching harmonics to the grid to meet statutory requirements, such as IEEE519-2014 [6] and IEC61727-2004 [7].

1.1.1.1 Switching Network

Based on input source types, the switching network of DG grid-connected converters can be classified into two basic types as shown in Figure 1.4 [8]. Figure 1.4a, b depict a two-level bridge for the current source converter and voltage source converter (VSC), respectively. The number n varies with the system configuration.

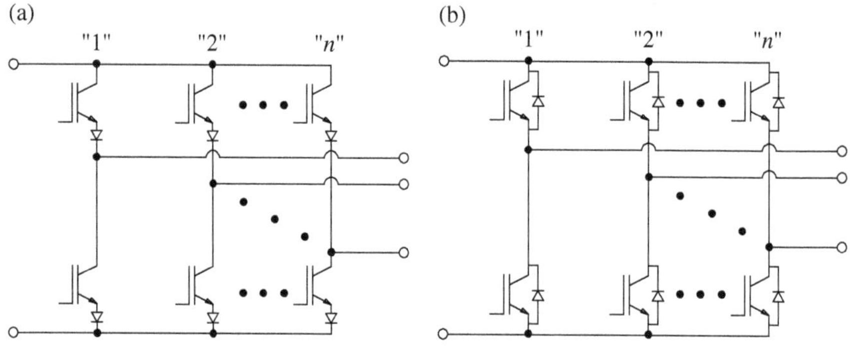

Figure 1.4 Configurations of switching networks. (a) Current source converter; (b) voltage source converter.

Considering that a VSC with bidirectional power feature plays a major role in distributed renewable generation systems, switching networks with $n = 1, 2$, and 3 for single- and three-phase voltage-source inverters (VSIs) are extensively studied.

1.1.1.2 Output Filter

The power grid is an infinite bus bar; hence, the grid interfacing circuit at the point-of-common-coupling (PCC) must be inductive. The simplest form of output filter for VSI is an inductor (L). The lower limit of the filter inductance is determined by the harmonic requirement of the grid-injected current according to IEEE 519-1992, where the percentage relative to the nominal grid-side fundamental current is specified in Table 1.1. The harmonic currents at different harmonic frequencies can be calculated from the ratio of corresponding harmonic voltage amplitudes to the filter impedance characteristics. The first-order L-type filter has a simple structure and is easy to install; however, it has relatively poor attenuation of switching harmonics, resulting in a very large size.

Thus, as shown in Figure 1.5, the industry prefers to use higher-order filters, such as a third-order inductor-capacitor-inductor (*LCL*) filter or a third-order

Table 1.1 Maximum permitted harmonic current distortion in percentage of current I_g according to IEEE 519-1992.

Individual harmonic order h	h<11	11≤h<17	17≤h<23	23≤h<35	35<h	THD
Percentage (%)	4.0	2.0	1.5	0.6	0.3	5.0

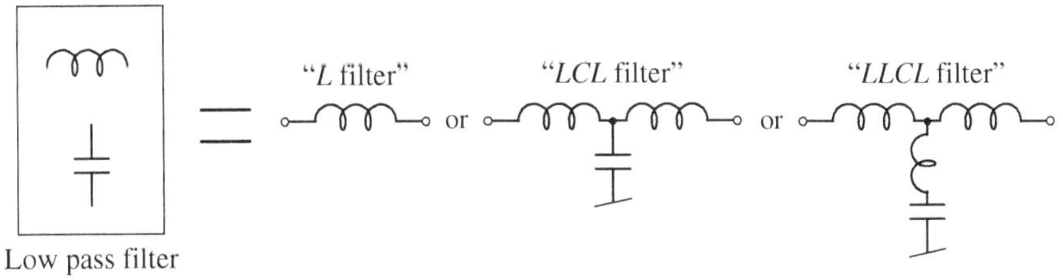

Figure 1.5 Filter configurations.

LCL filter with a parallel notch circuit (*LLCL* filter) [9] to achieve better switching harmonic attenuation with small reactive elements.

1.1.2 Control Stage

Depending on their operation, DG grid-connected converters can be classified into grid-feeding, grid-supporting, and grid-forming power converters [10, 11].

The grid-forming converters can be represented as an ideal AC voltage source with a low-output impedance by setting the voltage amplitude E^* and frequency ω^* of the local grid using a proper control loop as illustrated in Figure 1.6a.

Correspondingly, grid-feeding power converters are designed to deliver power to an energized grid. They can be represented as an ideal current source connected to the grid in parallel with high impedance. The simplified scheme of the grid-feeding power converter is depicted in Figure 1.6b where P^* and Q^* represent the active and reactive powers to be delivered, respectively. In this application, it is important to note that the current source should be perfectly synchronized with the AC voltage at the connection point to accurately regulate the active and reactive power exchanged with the grid. Currently, the grid-feeding power converter is still the mainstream of grid-connected converters, and it is also the focus of the following study.

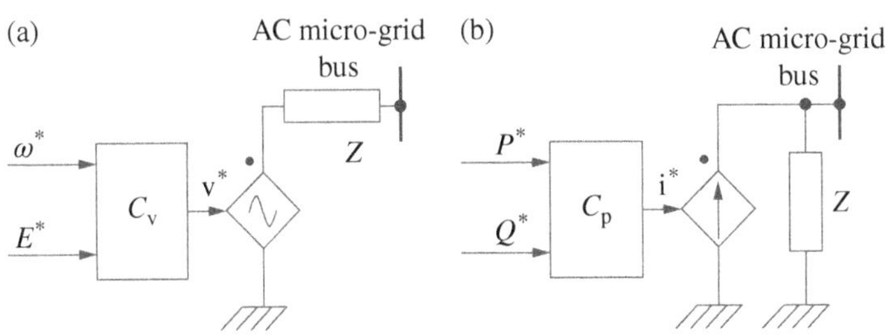

Figure 1.6 Simplified representation of grid-connected power converters. (a) Grid-forming; (b) grid-feeding.

1.2 Challenges for Controlling DG Grid-Connected VSCs with High-Order Power Filter

Modern DG grid-connected VSC must remain stable for different grid conditions, especially when higher-order output filters such as *LCL* or *LLCL* filters are used to minimize the size of reactive components. Additionally, they are necessary to address many intrinsic and extrinsic challenges in current regulation, such as oscillation due to filter resonance, filter parametric sensitivities, digital delay, uncertainty of the grid condition, and disturbances at the PCC.

1.2.1 Intrinsic Challenges

- Filter resonance

To reduce the size and weight of the filter, the most commonly applied solution is to use a high-order *LCL* or *LLCL* filter instead of a first-order *L* filter. However, the presence of multiple reactive elements, such as inductive and capacitive components, results in resonant frequencies. Thus, the output resonates when the inverter with *LCL* or *LLCL* filter is subject to disturbances [9]. In particular, filter resonance results in system instability [12, 13].

Depending on the output current detection positions as shown in Figure 1.7a, there are two methods for regulating the output current of the system: (i) by directly controlling the grid current or (ii) by controlling the converter-side current. The transfer function characteristics in the frequency domain are plotted in Figure 1.7b where the resonant poles are formed by the parallel converter grid-side inductors and filter capacitor, and the complex-conjugate zeros are formed by the grid-side inductor(s) and filter capacitor. If single-loop converter-side current regulation is chosen, the conjugate zeros aim to reduce the effect of resonant poles and improve system stability [14]. Hence, this type of regulation is commonly used in many practical applications owing to its inherent damping capability and overcurrent protection purposes [15, 16]. However, it is challenging to design a controller by considering both the variable resonant poles and conjugate zeros under variable grid conditions. Moreover, the converter-side current regulation does not directly offer harmonic rejection capability of grid current against grid harmonic voltage. If single-loop grid-side current regulation is chosen, the phase angle around the resonant poles decreases sharply and the resonant peak excites the potential oscillation around the filter resonant frequency, thus affecting the system stability. In digital control implementation, the sampling-and-computational delay time would be favorable in regulating the grid-side current owing to its inherent damping properties, which was reported in previous

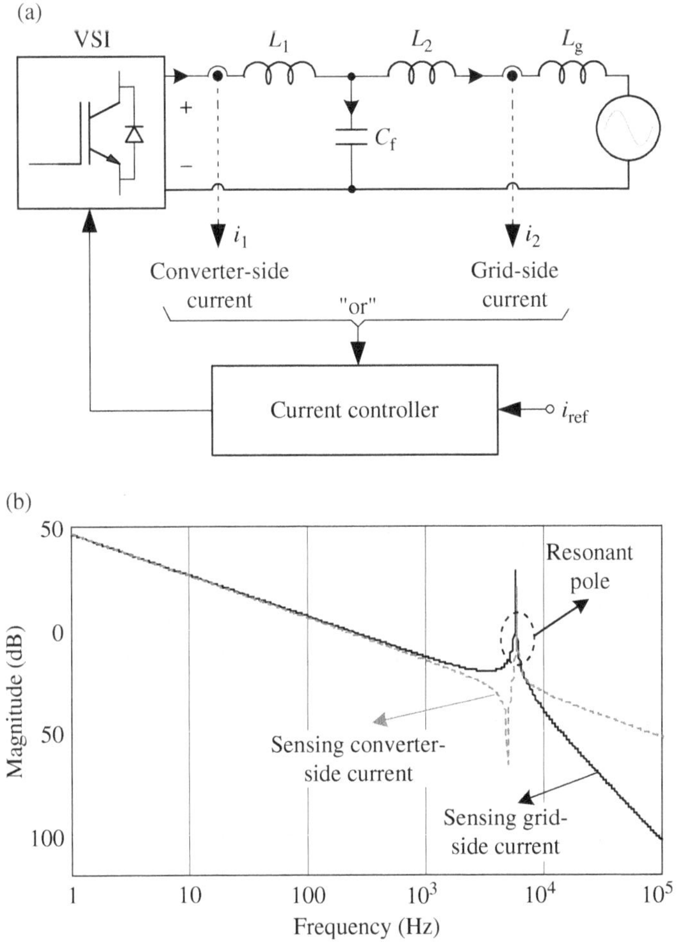

Figure 1.7 Different current regulation schemes and their characteristics. (a) Two main current regulation configurations; (b) comparison of open-loop transfer characteristics.

studies [17]. However, the filter resonant frequency must be higher than the critical frequency and one-sixth of the sampling frequency.

1.2.1.1 Filter Parametric Sensitivities

It is essential to design a system that is stable for specific filter parameters. However, the significant tolerance of filter parameters for example, ±20% variation of inductance and ±10% variation of capacitance still challenges the system stability. Thus, parametric sensitivities should be examined in the controller design.

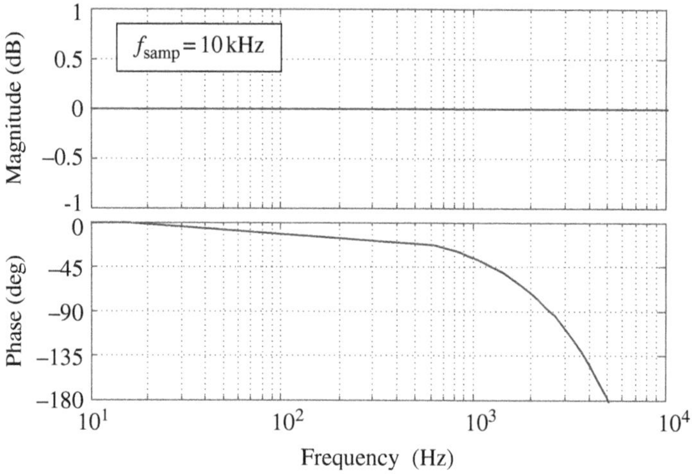

Figure 1.8 Significant phase delay with sampling delay.

1.2.1.2 Digital Delay

Digital delays, such as sampling and computational delays, are inherently present in digital control systems. This results in a significant phase delay as the frequency increases as shown in Figure 1.8. This adversely affects the system robustness and control bandwidth.

1.2.2 Extrinsic Challenges

1.2.2.1 Grid Impedance Variation

Owing to the growing integration of decentralized DG systems with long transmission cables and transformers, the non-negligible grid impedance poses challenges to circuit designers in developing an effective control scheme that can ensure the operational stability of inverters both in stiff and weak-grid conditions [18, 19]. This is because the filter resonant frequency varies with grid impedance as shown in Figure 1.9. Thus, impedance levels must be considered when installing inverters in large power plants or remote areas. For simplicity, the grid impedance level is usually denoted as a percentage of the base impedance value of the inverter or short-circuit ratio at the PCC between the short-circuit power of the AC power system and maximum apparent power of the shunted inverter [3].

1.2.2.2 Disturbances at the PCC

A pure sinusoidal grid voltage is produced by an ideal AC generator built with finely distributed stator and field windings that operate in a uniform magnetic field. However, nonlinear electronic loads in shunted grids cause the current to

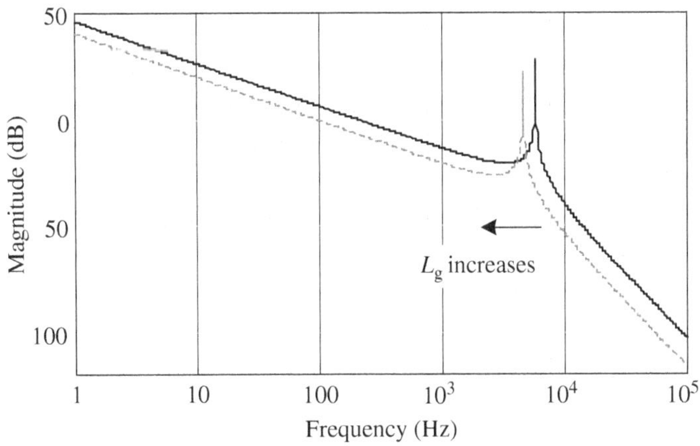

Figure 1.9 Transfer characteristics with grid impedance variation.

vary disproportionately with the grid voltage during each cyclic period. When there is significant impedance in the path from the power source to a nonlinear load as shown in Figure 1.10, these current distortions will also produce distortions or harmonics in the voltage waveform at the load [20, 21]. Thus, disturbances at the PCC caused by the source or nonlinear loads can sometimes affect the normal operation of the inverter. Conversely, inverters with slow dynamics or poor harmonic suppression deteriorate the voltage distortion at the PCC [22].

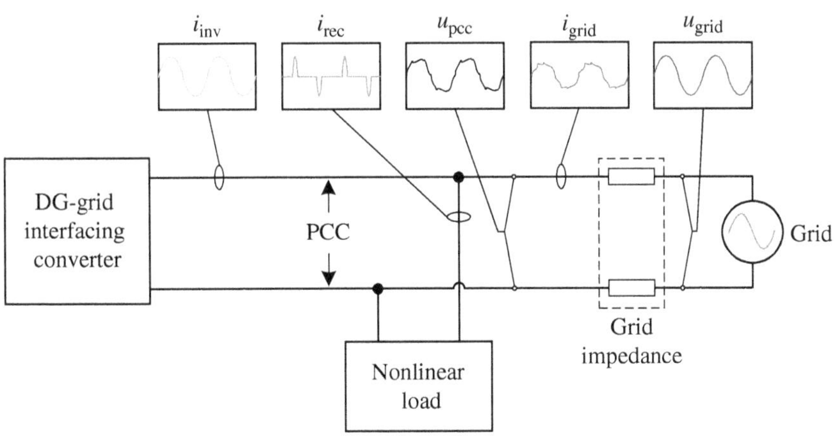

Figure 1.10 Disturbances caused by nonlinear loads at the PCC.

1.3 Methods for Controlling DG Grid-Connected VSCs with High-Order Power Filter

1.3.1 Methodologies to Assess the Stability of DG Grid-Connected VSCs

To address the abovementioned challenges, it is essential to develop effective modeling and analysis of the harmonic stability problem of DG grid-connected VSCs.

1.3.1.1 Eigenvalue-Based Analysis

Eigenvalue-based analysis is widely used to assess the small-signal stability of AC power systems [23, 24]. This analysis is usually based on the study of eigenvalues of a system state space model matrix and requires adequate information of the elements in the system, including the physical characteristics and control parameters [25]. However, the control parameters of grid-tied inverters are generally unavailable especially when analyzing the stability of a large-scale power-converter-based AC distributed power generation system (DPGS).

1.3.1.2 Impedance-Based Stability Analysis

Impedance-based stability analysis is based on generalized Nyquist stability criterion proposed by Middlebrook to assess the stability of the DC system [26]. It is used to analyze the stability of AC DPGS [22, 25–29].

A detailed description of impedance-based stability method for current-controlled grid-tied inverters that are widely used in DPGSs is given in [22]. As shown in Figure 1.11, a current-controlled inverter has a current source I_c in parallel with the output impedance Z_o under zero grid impedance condition, and the grid is represented by a voltage source V_g in series with the equivalent grid impedance Z_g. The grid impedance consists of line impedance and transformer output impedance; however, Z_g for a local current-controlled inverter still depends on the number of grid-tied inverters connected to the PCC [30].

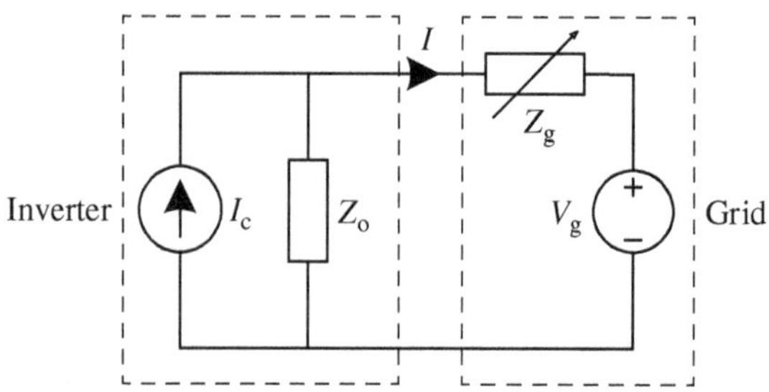

Figure 1.11 Equivalent circuit of a current-control-based grid-tied inverter.

Based on Thevenin and Norton equivalent circuits, if the grid voltage is assumed to be stable without the inverter and the inverter is assumed to be stable when the grid impedance is zero, the injected current from the grid-tied inverter is defined as follows:

$$I(s) = \left[I_c(s) - \frac{V_g(s)}{Z_o(s)}\right] \cdot \frac{1}{1 + Z_g(s)/Z_o(s)} \tag{1.1}$$

For the current-controlled grid-tied inverter system to be stable, the minor-loop gain $Z_g(s)/Z_o(s)$ must satisfy the generalized Nyquist stability criterion.

1.3.1.3 Application Issue Related to Impedance-Based Stability Analysis

The principle of impedance-based stability analysis is widely accepted by academic and industrial communities. However, many scholars are still divided in their opinion on the application of impedance-based stability analysis to the design of grid-tied power converters in AC DPGS.

Impedance-based stability analysis is a type of small-signal analysis method, and AC DPGS has no time-invariant operating point. Hence, researchers performed the analysis in a rotating (dq) reference frame. As a result, the minor-loop-gain matrix has four items: Z_{dd}, Z_{qq}, Z_{dq}, and Z_{qd}. In addition, different stability criteria were induced; however, a more convincing mathematical analysis and derivation of these stability criteria need further exploration. Moreover, an effective stability criterion in the dq frame has not been investigated for a three-phase unbalanced AC DPGS [31].

In addition to the dq frame, impedance-based stability analysis was performed in a stationary frame [22, 31]. From an engineering point of view, it can be noted that, as long as the switching frequency of the inverters is higher than the line frequency of 50/60 Hz, small-signal analysis can be simplified.

In fact, for a linear-control-based grid-tied inverter, system stability assessed by impedance-based stability analysis is approximately equal to that of the Bode diagram analysis. Impedance-based stability analysis focuses on the interaction between Z_g and Z_o of the local grid-tied inverter while Bode diagram analysis focuses on the stability of the entire system including Z_g. Unlike Bode diagram analysis, the impedance of the grid can be tested or estimated online using impedance-based stability analysis. However, universal quantitative stability indices for the AC system, such as phase margin and gain margin in Bode diagram analysis, have not been defined for impedance-based stability analysis.

Theoretically, impedance-based stability analysis should sample grid impedance information online. However, the measured grid impedance may not be reliable because of the possible harmonic voltages at the PCC. Currently, it is difficult to use impedance-based stability analysis to derive guidelines for designing the controller for converters quantitatively. Nevertheless, it is still a useful tool for qualitatively analyzing the enhancement of the stability of an AC DPGS. For example, as shown in Figure 1.2, Z_o, Z_g, or both Z_o and Z_g can be reshaped to ensure the stability of the local *LCL*-filter-based current-controlled grid-tied inverter thereby enhancing the stability of the AC DPGS.

1.3.2 Methods to Mitigate Filter Resonance

To mitigate the resonance caused by the high-order filter and improve the stability margin of the system, proper damping is mandatory. Damping can be achieved by either passive or active methods, such as online grid impedance estimation, inherent damping, passive damping, active damping, and hybrid damping.

1.3.2.1 Online Grid Impedance Estimation

Theoretically, as long as Z_g can be estimated accurately and periodically, the control parameters can be adjusted online to modify Z_o and thereby ensure system stability under any grid condition [18, 32]. To estimate the variation in grid parameters, many online estimation methods were proposed [31, 33–50]. Based on the mode of realization, these methods can be classified into active, passive, and quasi-passive methods.

Active estimation methods, including noncharacteristic harmonic injection [33], broadband signal injection [34], *LCL* resonance excitation [35], and reactive power variation method [36, 37] deliberately disturb the grid and extract feedback information to estimate the grid impedance.

Passive estimation methods [31, 38–46] are based on direct measurement and evaluation of grid information using least-squares estimation algorithm [40], successive approximation estimation [38], inherent switching feature application [39], and plant time constant identification algorithm [40].

Quasi-passive estimation methods [47, 48] combine the nonintrusive advantage of a passive estimation method and accuracy of an active estimation method. In this method, passive estimation is first selected to estimate the grid impedance. When the value exceeds a selected evaluation index, the quasi-passive method switches to active estimation online.

The accuracy of grid impedance estimation method is inevitably affected by the harmonics generated by other grid-tied inverters at the same PCC. However, the accuracy, detection speed, and anti-interference ability of the online grid impedance estimation method must be improved. Currently, the grid impedance estimation method cannot be independently applied to grid-tied inverters, but it can be adopted as an additional method for enhancing their stability [49, 50].

1.3.2.2 Inherent Damping

The digital time delay induces inherent damping (ID) effect in the current control loop [14, 17, 51–54]. ID effect depends on the location of sensing of current and set time delay. Yin et al. [17] studied the damping characteristic of time delay in a grid-current-feedback-based single-loop control system, and many other studies [51–53] discussed the ID region of a grid-tied inverter. Wang et al. [54] summarized the previous research and derived the time delay range using different current feedback methods to improve the stability of a single-loop control system. As presented in [54], when inverter-side-inductor current feedback is used, the related time delay should satisfy the following condition:

$$\begin{cases} T_\mathrm{d} < \dfrac{1}{4f_\mathrm{r}} (k = 0) \\ \dfrac{4k-1}{4f_\mathrm{r}} < T_\mathrm{d} < \dfrac{4k+1}{4f_\mathrm{r}} (k = 1, 2, \ldots) \end{cases} \tag{1.2}$$

where T_d is the total time delay and $f_\mathrm{r} = [(L_1 + L_2 + L_\mathrm{g})/(L_1(L_2 + L_\mathrm{g})C]^{1/2}/2\pi$ is the filter resonance frequency (L_1 is the inverter-side inductor, L_2 is the grid-side inductor, L_g is the grid inductor, and C is the filter capacitor).

When the grid-side-inductor current feedback is used, the related time delay should satisfy the following condition:

$$\frac{4k+1}{4f_\mathrm{r}} < T_\mathrm{d} < \frac{4k+3}{4f_\mathrm{r}} (k = 0, 1, 2, \ldots), \tag{1.3}$$

Based on the ongoing research, two key points must be considered when using the ID technique:

1) When Z_g changes, f_r cannot shift across the critical frequency, which is defined as:

$$f_\mathrm{crit} = 1/4\tau T_\mathrm{s} \tag{1.4}$$

   where τT_s is the total delay, including the computational delay and pulse-width modulation (PWM) delay.
2) If the cable is long enough, the system may have multi-resonance peaks where gains are greater than 0 dB owing to the distributed capacitances [55, 56]. Therefore, Eqs. (1.2) and (1.3) should be satisfied at every resonant frequency.

In addition, it is difficult to apply the ID technique to a current-controlled inverter independently when Z_g varies and multiple resonance peaks exist.

1.3.2.3 Passive Damping

Passive damping (PD) is easy to implement and widely used in industrial applications [57–61]. The parameter design of PD is based on the physical characteristics

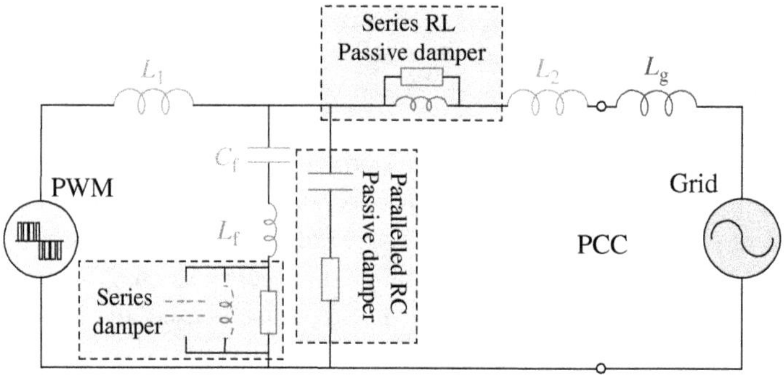

Figure 1.12 Possible passive damping methods for a grid-tied inverter.

of the system, such as resistive region [58] and Q-factor [59]. Figure 1.10 illustrates the different types of PD where L_f is the trap inductor for the *LLCL* filter. As shown in Figure 1.12, the insertion of a resistor(s) to damp the possible resonance results in extra power losses.

Figure 1.13 shows the open-loop Bode diagram of $i_g(s)/i_{gref}(s)$ for a grid-tied inverter with only *RC* passive damping [32, 59]. It can be observed that a single

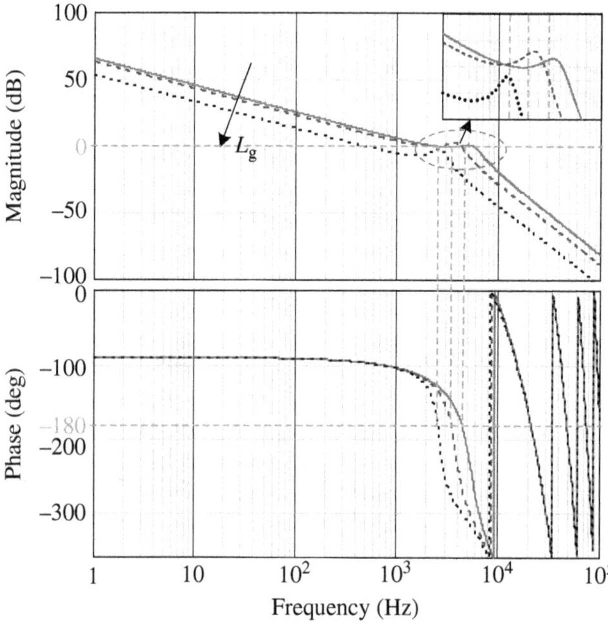

Figure 1.13 Open-loop gain $i_g(s)/i_{gref}(s)$ of the *LCL*-filter-based grid-tied inverter with *RC* damper.

PD-based system cannot achieve high bandwidth in a weak grid. However, a composite PD method can handle the variation in Z_g at the cost of additional materials and increased damping power losses [60].

1.3.2.4 Active Damping

Active damping (AD) is a unique method in state feedback-based control. In contrast to PD, AD [51, 62–80]has received more attention over the years because of its flexibility and decreased damping power losses; however, it increases the control complexity and cost of the sensors [63]. There are two main types of AD: single loop and multiloop. Figure 1.14 shows the control block of an *LCL*-filter-based grid-tied inverter with a digital filter and AD with capacitor current feedback where $G_h(s)$ is the zero-order-hold delay, $G_d(s)$ is the computational delay, $1/T_s$ is the sampler, T_s is the sampling period, G_{inv} is the inverter gain, K_{ic} and K_{ig} are the feedback constants of the capacitor current and grid current, respectively, $G_c(s)$ is the proportional controller, and $H_{dig}(s)$ is the digital filter.

As shown in Figure 1.14, when $K_{ic} = 0$, a single-loop control AD is obtained. The digital filter $H_{dig}(s)$ used in single-loop control consists of a notch filter, lead-lag network, and bi-quad filter [64]. Alzola et al. [65] proposed a self-commissioning notch filter to deal with the variation in Z_g. However, this method still relies on the accuracy of the real-time resonance frequency estimation, which is inevitably affected by harmonics at the PCC. Yao et al. [66] suggested a digital notch-filter-based AD technique in which the notch filter frequency is different from the nominal resonant frequency of the *LCL* filter. Nevertheless, if multi-resonance peaks appear, this damping method may become ineffective.

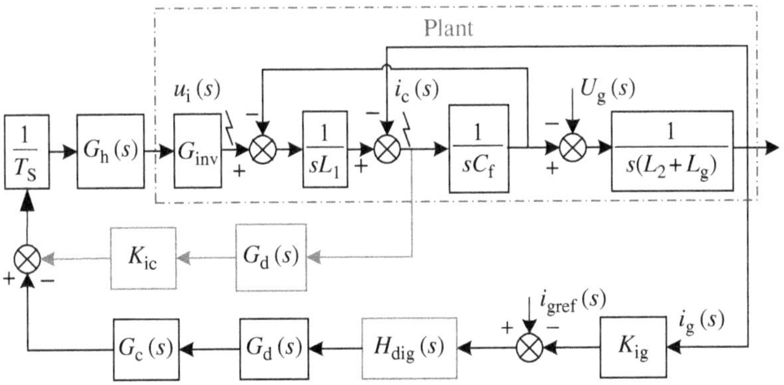

Figure 1.14 Control diagram of *LCL*-filter-based grid-tied inverter using a digital filter and AD method with capacitor current feedback.

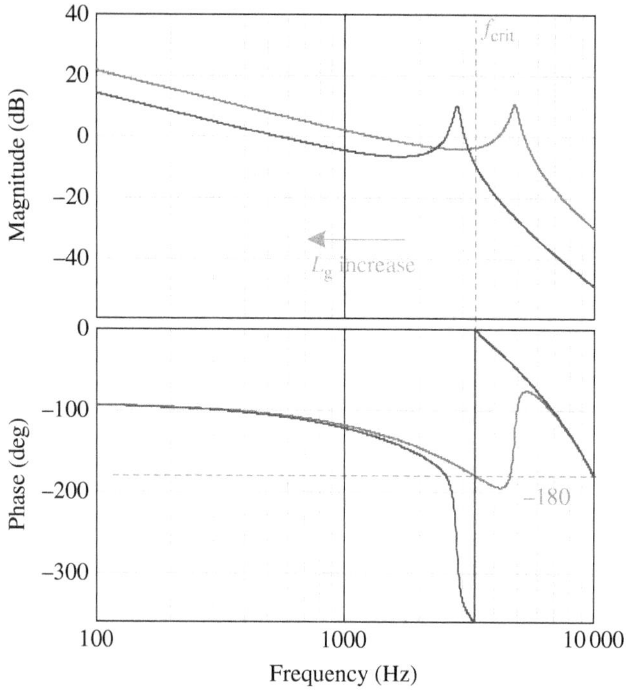

Figure 1.15 Open-loop gain $i_g(s)/i_{gref}(s)$ of system using capacitor current feedback damping method.

Multiloop AD methods include a virtual resistor [67–69], capacitor current feedback [70–75], capacitor voltage feedback [76–79], and *LC*-trap voltage feedback [80].

Figure 1.15 depicts the open-loop gain $i_g(s)/i_{gref}(s)$ of a current-controlled inverter using capacitor current feedback damping method only. When the characteristic frequency of the system is greater than f_{crit}, the system exhibits a non-minimum-phase response as a result of the appearance of right-half-plane poles, thus complicating the controller design.

Many different methods have been proposed for modifying the capacitor current feedback AD. Yang et al. [72] proposed a real-time computational method to reduce the computational delay, which can simplify the closed-loop design and improve the system performance effectively. However, when the grid voltage has many harmonics or many other factors exist that result in a sharp variation of duty cycle of the grid-tied inverter, this method will not work efficiently. Wang et al. [67] and Li et al. [74] aimed to increase f_{crit} from $1/6T_s$ to $1/4T_s$ or $1/3T_s$ by modifying the feedback loop where the damping region can be widened. However, the high-pass filter inserted into the capacitor current feedback inner loop [67] increases the high-frequency noise interference and the proposed optimal damping coefficient in [74] cannot be derived precisely.

To ensure the stability of the system when f_r changes across f_{crit}, Pan et al. [70] modified the capacitor current feedback AD by inserting a phase-lag component in the control loop. With a digital delay of $0.75T_s$, this method can maintain system stability when Z_g varies widely. However, the phase margin of the system is not sufficiently large when the characteristic frequency is close to f_{crit}. To ensure a sufficient phase margin, Bao et al. [73] proposed the forbidden frequency region of $[(1-\xi)f_{crit}, (1+\xi)f_{crit}]$, where ξ is the coefficient of the forbidden region. Nevertheless, when ω_{res} is strictly less than $(1-\xi)f_{crit}$, it is difficult to maintain high bandwidth of the system in a weak grid. Additionally, when f_r is strictly within the range of $[(1+\xi)f_{crit}, 1/2T_s]$, it is difficult to maintain a good phase margin in the parametric design of *LCL* filter when Z_g varies significantly.

Similar to the capacitor current feedback AD method, the derivative feedback of capacitor voltage can also induce effective AD [76–78]. Using an *LLCL* filter, Huang et al. [80] compared the capacitor current feedback AD with the *LC*-trap voltage feedback AD and found that the latter had a wider damping region. However, as the grid voltage experiences a sudden and large change, it is difficult to choose a suitable feedback coefficient to ensure a satisfactory damping effect under different grid voltage conditions.

Furthermore, similar to the ID method, if multi-resonance peaks appear owing to the distributed capacitance of the cable, AD method (single-loop or multiloop) may not be effective especially when Z_g varies over a wide range and the grid background harmonics are abundant.

1.3.2.5 Hybrid Damping

Hybrid damping (HD) techniques have been discussed in previous research works [32, 81, 82]. These techniques result in a trade-off between the power losses and damping robustness of the system. Lei et al. [81] proposed a hybrid damper that uses a resistor in series with a filter capacitor along with capacitor current feedback control where the extra damping losses may be high. The *RC* damper is preferred in HD because the extra power losses are negligible [32, 61, 82].

Liu et al. [32] combined an *RC* damper with a digital notch filter to dampen the resonance caused by an *LLCL* filter. The control diagram is shown in Figure 1.16.

$$Z_c(s) = \frac{(R_d C_d s + 1)(L_f C_f s^2 + 1)}{L_f C_f C_d s^3 + R_d C_d C_f s^2 + (C_d + C_f)s}, \tag{1.5}$$

where R_d and C_d are the parallel PD resistor and capacitor, respectively. A high control bandwidth can be achieved when Z_g varies widely. Therefore, a

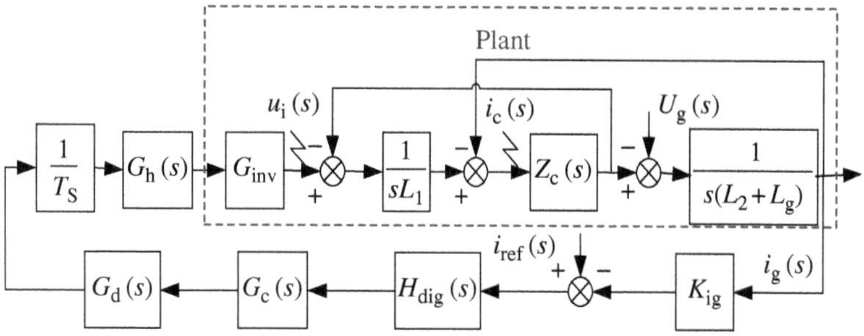

Figure 1.16 Control diagram of the hybrid damper composed of PD of *RC* and AD of the digital filter. *Source:* [32]/IEEE.

proportional plus resonant controller can be used to attenuate the low-frequency grid background harmonics.

1.3.3 Harmonic distortion Mitigation Methods

To mitigate the disturbances at the PCC caused by the source or nonlinear loads, there are three basic ways to suppress the effect of background harmonic voltage (BHV):

1) First is the grid-voltage feed-forward method, which is widely used to suppress the low-order harmonics of the grid-injected current [83–85]. For example, [9] adopted an adaptive feed-forward grid voltage method to obtain an appropriate phase margin or high control bandwidth while the equivalent grid impedance changes. However, the reliability of this method still relies on the accuracy of detection of the feed-forward grid voltage and estimated grid impedance; however, it is difficult to maintain reliability in a real grid full of high-order BHVs.
2) Second is to insert a proportional resonant regulator with a harmonic compensator (PR + HC regulator) for a grid-tied inverter using a low control bandwidth [86] where the lowest frequency of HC is higher than the control bandwidth under any grid condition. The compensated gain at the high-order harmonic frequencies should be large enough to suppress the effect of BHV, thus making the grid-tied inverter susceptible to the frequency drift of the grid.
3) Third is to adopt a PR + HC regulator for a grid-tied inverter based on a high-bandwidth control strategy [18, 87, 88] as presented in Figure 1.17a. In Figure 1.17b, the frequency of the highest HC should be within the control bandwidth under any grid condition to ensure the overall stability of the system. In this case, the control bandwidth of the grid-tied inverter is expected to be as high as possible in the weakest grid state.

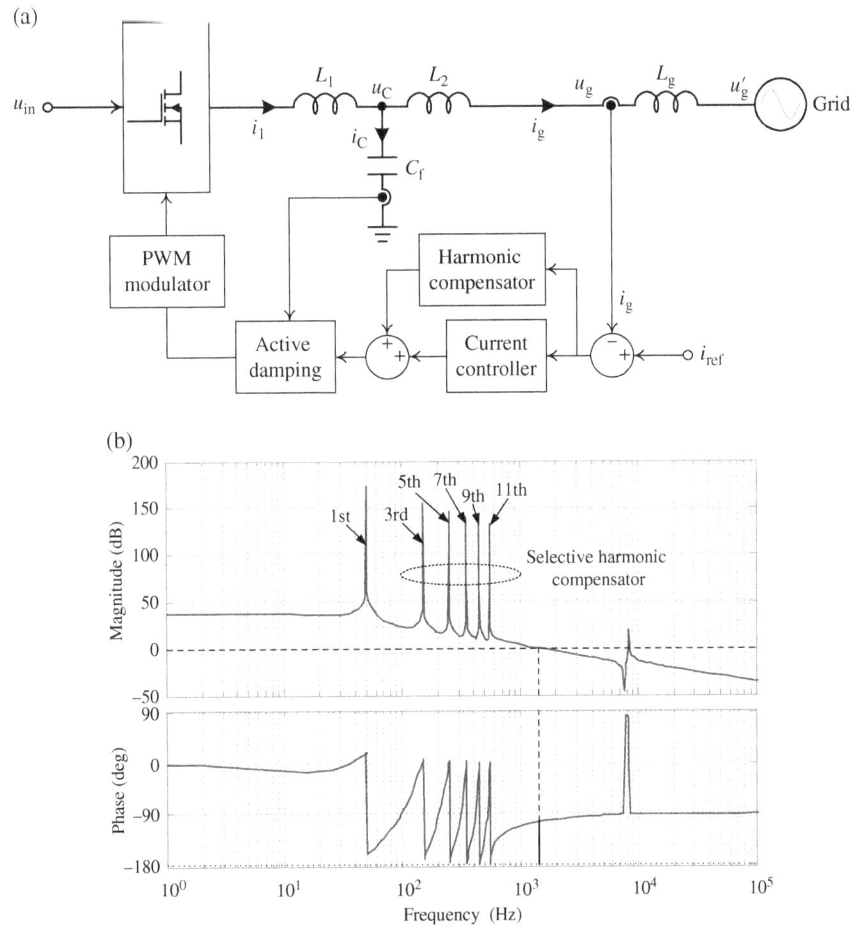

Figure 1.17 Control characteristics with selective harmonic compensator. (a) Control architecture; (b) open-loop gain.

1.4 Supplementary Note

In terms of topology, the *LCL*/*LLCL* filter design method introduced in this study can be used for single-phase inverters and three-phase four-wire grid-connected inverters. A similar idea of filter design can be used for three-phase three-wire grid-connected inverter circuits. The switching order and double-frequency switching order harmonics in the output voltage spectrum of a three-phase three-wire inverter with PWM modulation are relatively abundant. Therefore, compared to the *LCL* filter, the advantage of *LLCL* filter in saving inductance is not obvious. Note: In the future, a three-phase three-wire grid-connected inverter system with a new filter topology scheme can be developed.

Additionally, the performance of the controller improves when the control of the grid-connected inverter gradually shifts from classic linear control [89] to nonlinear control [90–93] and finally develops into intelligent control. This study introduces the concept of classic linear control methods that are limited to the length and focus of this study. The nonlinear control of grid-connected inverters and intelligent control methods will be introduced in the future.

References

1 Renewables Global Status Report 2020 Update. http://www.ren21.net/ (accessed June 2020).

2 Xue, Y., Chang, L., Kjaer, S.B. et al. (2004). Topologies of single phase inverters for small distributed power generators: an overview. *IEEE Trans. Power Electron.* 19 (5): 1305–1314.

3 Blaabjerg, F., Teodorescu, R., Liserre, M., and Timbus, A.V. (2006). Overview of control and grid synchronization for distributed power generation systems. *IEEE Trans. Ind. Electron.* 53 (5): 1398–1409.

4 Teodorescu, R., Liserre, M., and Rodríguez, P. (2011). *Grid Converters for Photovoltaic and Wind Power Systems.* Hoboken: Wiley.

5 Guerrero, J.M., Vasquez, J.C., Matas, J. et al. (2011). Hierarchical control of droop-controlled AC and DC microgrids – a general approach toward standardization. *IEEE Trans. Ind. Electron.* 58 (1): 158–172.

6 IEEE (2014). *519-2014 – IEEE Recommended Practice and Requirements for Harmonic Control in Electric Power Systems.* IEEE.

7 IEC (2004). *61727:2004 – Photovoltaic (PV) Systems – Characteristics of the Utility Interface.* IX-IEC.

8 Azmi, S.A., Ahmed, K.H., Finney, S.J., and Williams, B.W. (2011). Comparative analysis between voltage and current source inverters in grid-connected application. *IET Conference on RPG*, pp. 1–6, Edinburgh (September 2011). London: IFT.

9 Wu, W., He, Y., and Blaabjerg, F. (2012). An LLCL power filter for single-phase grid-tied inverter. *IEEE Trans. Power Electron.* 27 (2): 782–789.

10 Engler, A. (2004). Control of inverters in isolated and in grid tied operation with regard to expandability in tutorial: power electronics for regenerative energy. *Proceedings of the IEEE Power Electronics Specialists Conference*, Aachen, Germany (20–25 June 2004).

11 De Brabandere, K., Bolsens, B., Van den Keybus, J. et al. (2007). A voltage and frequency droop control method for parallel inverters. *IEEE Trans. Power Electron.* 22 (4): 1107–1115.

12 Zue, A.O. and Chandra, A. (2006). Simulation and stability analysis of a 100 kW grid connected LCL photovoltaic inverter for industry. *Proceedings of the Power Energy Society General Meeting*, p. 6, Montreal, QC (June 2006). New York: IEEE.

13 Blasko, V. and Kaura, V. (1997). A novel control to actively damp resonance in input LC filter of a three-phase voltage source converter. *IEEE Trans. Ind. Appl.* 33 (2): 542–550.

14 Tang, Y., Loh, P.C., Wang, P. et al. (2012). Exploring inherent damping characteristic of LCL-filters for three-phase grid-connected voltage source inverters. *IEEE Trans. Power Electron.* 27 (3): 1433–1443.

15 Liserre, M., Blaabjerg, F., and Hansen, S. (2001). Design and control of an LCL filter based three-phase active rectifier. *Proceedings of the IEEE 36th IAS Conference*, pp. 299–307 (30 September—4 October 2001). New York: IEEE.

16 Teodorescu, R., Blaabjerg, F., Borup, U., and Liserre, M. (2004). A new control structure for grid-connected LCL PV inverters with zero steady-state error and selective harmonic compensation. *Proceedings of the IEEE 19th APEC*, pp. 580–586, Anaheim, CA (February 2004). New York: IEEE.

17 Yin, J., Duan, S., and Liu, B. (2013). Stability analysis of grid-connected inverter with LCL filter adopting a digital single-loop controller with inherent damping characteristic. *IEEE Trans. Ind. Inf.* 9 (2): 1104–1112.

18 Liserre, M., Teodorescu, R., and Blaabjerg, F. (2006). Stability of photovoltaic and wind turbine grid-connected inverters for a large set of grid impedance values. *IEEE Trans. Power Electron.* 21 (1): 263–272.

19 Chen, X., Gong, C.Y., Wang, H.Z., and Cheng, L. (2012). Stability analysis of LCL-type grid-connected inverter in weak grid systems. *Proceedings of the IEEE Renewable Energy Research and Applications Conference*, pp. 1–6, Nagasaki, Japan (November 2012). New York: IEEE.

20 Yao, Z. and Xiao, L. (2013). Control of single-phase grid-connected inverters with nonlinear loads. *IEEE Trans. Ind. Electron.* 60 (4): 1384–1389.

21 Mazumder, S.K., Burra, R.K., Huang, R. et al. (2010). A universal grid-connected fuel-cell inverter for residential application. *IEEE Trans. Ind. Electron.* 57 (10): 3431–3447.

22 Sun, J. (2011). Impedance-based stability criterion for grid-connected inverters. *IEEE Trans. Power Electron.* 26 (11): 3075–3078.

23 Kundur, P. (1994). *Power System Stability and Control.* New York: *McGraw-Hill.*

24 Ilic, M. and Zaborszky, J. (2000). *Dynamics and Control of Large Electric Power Systems.* New York: *Wiley.*

25 Belkhayat, M. (1997). Stability criteria for ac power systems with regulated loads. PhD dissertation. Purdue University, West Lafayette.

26 Middlebrook, R.D. (1976). Input filter considerations in design and application of switching regulators. *Record of 1976 IEEE Industry Applications Society Annual Meeting*, pp. 366–382, Chicago, IL. New York: IEEE.

27 Mao, H., Boroyevich, D., and Lee, F.C. (1998). Novel reduced-order small signal model of a three-phase PWM rectifier and its application in control design and system analysis. *IEEE Trans. Power Electron.* 13 (3): 511–521.

28 Wen, B. (2014). Stability analysis of three-phase AC power systems based on measured D-Q frame impedances. PhD dissertation. Virginia Polytechnic Institution and State University, Blacksburg, VA.

29 Liu, F., Liu, J., Zhang, H., and Xue, D. (2014). Stability issues of Z+Z type cascade system in hybrid energy storage system (HESS). *IEEE Trans. Power Electron.* 29 (11): 5846–5859.

30 Agorreta, J.L., Borrega, M., Lopez, J., and Marroyo, L. (2011). Modeling and control of N-paralleled grid-connected inverters with LCL filter coupled due to grid impedance in PV plants. *IEEE Trans. Power Electron.* 26 (3): 770–785.

31 Sun, J., Bing, Z., and Karimi, K.J. (2009). Input impedance modeling of multipulse rectifiers by harmonic linearization. *IEEE Trans. Power Electron.* 24 (12): 2812–2820.

32 Liu, Y., Wu, W., He, Y. et al. (2016). An efficient and robust hybrid damper for LCL- or LLCL-based grid-tied inverter with strong grid-side harmonic voltage effect rejection. *IEEE Trans. Ind. Electron.* 63 (2): 926–936.

33 Timbus, A.V., Teodorescu, R., Blaabjerg, F., and Borup, U. (2004). Online grid measurement and ENS detection for PV inverter running on highly inductive grid. *IEEE Trans. Power Electron.* 2 (3): 77–82.

34 Roinila, T., Vilkko, M., and Sun, J. (2014). Online grid impedance measurement using discrete-interval binary sequence injection. *IEEE J. Emerging Sel. Topics Power Electron.* 2 (4): 985–993.

35 Liserre, M., Blaabjerg, F., and Teodorescu, R. (2007). Grid impedance estimation via excitation of LCL-filter resonance. *IEEE Trans. Ind. Appl.* 43 (5): 1401–1407.

36 Timbus, A.V., Rodriguez, P., Teodorescu, R., and Ciobotaru, M. (2007). Line impedance estimation using active and reactive power variations. *Proceedings of the IEEE PESC*, pp. 1273–1279, Orlando, FL (17–21 June 2007). IEEE.

37 Ciobotaru, M., Teodorescu, R., Rodriguez, P. et al. (2007). Online grid impedance estimation for single-phase grid-connected systems using PQ variations. *Proceedings of the IEEE PESC*, pp. 2306–2312, Orlando, FL (17–21 June 2007). IEEE.

38 Cobreces, S., Bueno, E.J., Pizarro, D. et al. (2009). Grid impedance monitoring system for distributed power generation electronic interfaces. *IEEE Trans. Instrum. Meas.* 58 (9): 3112–3121.

39 Cobreces, S., Rodriguez, P., Pizarro, D. et al. (2007). Complex-space recursive least squares power system identification. *Proceedings of PESC*, pp. 2478–2484, Orlando, FL (17–21 June 2007). IEEE.

40 Ciobotaru, M., Agelidis, V., and Teodorescu, R. (2011). Line impedance estimation using model based identification technique. *Proceedings of the IEEE EPE2011*, pp. 1–9, Birmingham, UK (30 August–1 September 2011). IEEE.
41 Gu, H., Guo, X., Wang, D., and Wu, W. (2012). Real-time grid impedance estimation technique for grid-connected power inverters. *Proceedings of the IEEE ISIE2012*, pp. 1621–1626, Hangzhou, China (28–31 May 2012). IEEE
42 Vidal, A., Yepes, A.G., Freijedo, F.D. et al. (2015). A method for identification of the equivalent inductance and resistance in the plant model of current-controlled grid-tied inverters. *IEEE Trans. Power Electron.* 30 (12): 7245–7261.
43 Pedersen, K.O.H., Nielsen, A.H., and Poulsen, N.K. (2003). Short-circuit impedance measurement. *IEE Proc.-Gener. Transm. Distrib.* 150: 169–174.
44 Bien, A., Borkowski, D., and Wetula, A. (2007). Estimation of power system parameters based on load variance observations – laboratory studies. *Proceedings of EPQU*, pp. 1–6, Barcelona, Spain (9–11 October 2007). IEEE.
45 Hoffmann, N. and Fuchs, F. (2014). Minimal invasive equivalent grid impedance estimation in inductive-resistive power networks using extended Kalman filter. *IEEE Trans. Power Electron.* 29 (2): 631–641.
46 Beres, R.N., Wang, X., Blaabjerg, F. et al. (2016). Optimal design of high-order passive-damped filters for grid-connected applications. *IEEE Trans. Power Electron.* 31 (3): 2083–2098.
47 Ghorbal, M.J.-B., Ghzaiel, W., Slama-Belkhodja, I., and Guerrero, J.M. (2012). Online detection and estimation of grid impedance variation for distributed power generation. *Proceedings of the IEEE MELECON*, pp. 555–560, Yasmine Hammamet, Tunisie (25–28 March 2012). IEEE.
48 Berriri, H., Naouar, M.W., and Slama-Belkhodja, I. (2012). Easy and fast sensor fault detection and isolation algorithm for electrical drives. *IEEE Trans. Power Electron.* 27 (2): 490–499.
49 He, Y., Chung, H., Ho, C.N.M., and Wu, W. (2017). Direct current tracking using boundary control with second-order switching surface for three-phase three-wire grid-connected inverter. *IEEE Trans. Power Electron.* 32 (7): 5723–5740.
50 He, Y., Chung, H.S.H., Ho, C.N.M., and Wu, W. (2017). Modified cascaded boundary-deadbeat control for a virtually-grounded three-phase grid-connected inverter with LCL filter. *IEEE Trans. Power Electron.* 32 (10): 5723–5740.
51 Parker, S.G., McGrath, B.P., and Holmes, D.G. (2014). Regions of active damping control for LCL filters. *IEEE Trans. Ind. Appl.* 50 (1): 424–432.
52 Zou, C., Liu, B., Duan, S., and Li, R. (2014). Influence of delay on system stability and delay optimization of grid-connected inverters with LCL filter. *IEEE Trans. Ind. Inf.* 10 (3): 1775–1784.
53 Tang, Y., Yoon, C., Zhu, R., and Blaabjerg, F. (2015). Generalized stability regions of current control for LCL-filtered grid-connected inverters without passive or active

damping. *Proceedings of IEEE Energy Conversion Congress and Exposition*, pp. 2040–2047, Montreal, Canada. IEEE.

54 Wang, J., Yan, J., Jiang, L., and Zou, J. (2016). Delay-dependent stability of single-loop controlled grid-connected inverters with LCL filters. *IEEE Trans. Power Electron.* 31 (1): 743–757.

55 Zhang, S., Jiang, S., Lu, X. et al. (2014). Resonance issues and damping techniques for grid-connected inverters with long transmission cable. *IEEE Trans. Power Electron.* 29 (1): 110–120.

56 Jiang, Y., Wu, W., He, Y. et al. (2017). New passive filter design method for overvoltage suppression and bearing currents mitigation in a long cable based PWM inverter-fed motor drive system. *IEEE Trans. Power Electron.* 32 (10): 7882–7893.

57 Beres, R., Wang, X., Blaabjerg, F. et al. (2016). A review of passive power filters for three-phase grid-connected voltage-source inverters. *IEEE Trans. Power Electron.* 4 (1): 54–69.

58 Balasubramanian, A.K. and John, V. (2013). Analysis and design of split-capacitor resistive inductive passive damping for LCL filters in grid-connected inverters. *IET Power Electron.* 6 (9): 1822–1832.

59 Wu, W., He, Y., Tang, T., and Blaabjerg, F. (2013). A new design method for the passive damped LCL- and LLCL-filter based single-phase grid-tied inverter. *IEEE Trans. Ind. Electron.* 60 (10): 4339–4350.

60 Wu, W., Sun, Y.J., Huang, M. et al. (2014). A robust passive damping method for LLCL-filter-based grid-tied inverters to minimize the effect of grid harmonic voltages. *IEEE Trans. Power Electron.* 29 (7): 3279–3289.

61 Alzola, R.P., Liserre, M., Blaabjerg, F. et al. (2013). Analysis of the passive damping losses in LCL-filter-based grid inverters. *IEEE Trans. Power Electron.* 28 (6): 2642–2646.

62 Hanif, M., Khadkikar, V., Xiao, W., and Kirtley, J.L. (2014). Two degrees of freedom active damping technique for filter-based grid connected PV systems. *IEEE Trans. Ind. Electron.* 61 (6): 2795–2803.

63 Bahrani, B., Vasiladiotis, M., and Rufer, A. (2014). High-order vector control of grid-connected voltage-source inverters with LCL-filters. *IEEE Trans. Ind. Electron.* 61 (6): 2767–2775.

64 Dannehl, J., Liserre, M., and Fuchs, F. (2011). Filter-based active damping of voltage source inverters with LCL filters. *IEEE Trans. Ind. Electron.* 58 (8): 3623–3633.

65 Alzola, R.P., Liserre, M., Blaabjerg, F. et al. (2014). Self-commissioning notch filter for active damping in three phase LCL-filter based grid inverters. *IEEE Trans. Power Electron.* 29 (12): 6754–6761.

66 Yao, W., Yang, Y., Zhang, X. et al. (2017). Design and analysis of robust active damping for LCL filters using digital notch filters. *IEEE Trans. Power Electron.* 32 (3): 2360–2375.

67 Wang, X., Blaabjerg, F., and Loh, P.C. (2010). Virtual RC damping of LCL-filtered voltage source inverters with extended selective harmonic compensation. *IEEE Trans. Power Electron.* 30 (9): 4726–4737.

68 Wang, X., Blaabjerg, X.F., and Loh, P.C. (2016). Grid-current-feedback active damping for LCL resonance in grid-connected voltage source inverters. *IEEE Trans. Power Electron.* 3 (1): 213–223.

69 Pan, D., Ruan, X., Bao, C. et al. (2014). Capacitor-current-feedback active damping with reduced computation delay for improving robustness of LCL-type grid-connected inverter. *IEEE Trans. Power Electron.* 29 (7): 3414–3427.

70 Pan, D., Ruan, X., Bao, C. et al. (2015). Optimized controller design for LCL-type grid-connected inverter to achieve high robustness against grid-impedance variation. *IEEE Trans. Ind. Electron.* 62 (3): 1537–1547.

71 Liu, F., Zhou, Y., Duan, S. et al. (2009). Parameter design of a two-current-loop controller used in a grid-connected inverter system with LCL filter. *IEEE Trans. Ind. Electron.* 56 (11): 4483–4491.

72 Yang, D., Ruan, X., and Wu, H. (2015). A real-time computation method with dual sampling modes to improve the current control performances of the LCL-type grid-connected inverter. *IEEE Trans. Ind. Electron.* 62 (7): 4563–4572.

73 Bao, C., Ruan, X., Wang, X. et al. (2014). Step-by-step controller design for LCL-type grid-connected inverter with capacitor-current-feedback active-damping. *IEEE Trans. Power Electron.* 29 (3): 1239–1253.

74 Li, X., Wu, X., Geng, Y. et al. (2015). Wide damping region for LCL-type grid-connected inverter with an improved capacitor-current-feedback method. *IEEE Trans. Power Electron.* 30 (9): 5247–5259.

75 Kukkola, J. and Hinkkanen, M. (2016). State observer for grid-voltage sensorless control of a inverter equipped with an LCL filter: direct discrete-time design. *IEEE Trans. Ind. Appl.* 52 (4): 3133–3145.

76 Dannehl, J., Fuchs, F.W., Hansen, S., and Thogersen, P.B. (2010). Investigation of active damping approaches for PI-based current control of grid-connected pulse width modulation inverters with LCL filters. *IEEE Trans. Ind. Appl.* 46 (4): 1509–1517.

77 Alzola, R.P., Liserre, M., Blaabjerg, F. et al. (2014). Systematic design of the lead-lag network method for active damping in LCL-filter based three-phase inverters. *IEEE Trans. Ind. Inf.* 10 (1): 43–52.

78 Xin, Z., Loh, P.C., Wang, X. et al. (2016). Highly accurate derivatives for LCL-filtered grid inverter with capacitor voltage active damping. *IEEE Trans. Power Electron.* 31 (5): 3612–3625.

79 Komurcugil, H., Altin, N., Ozdemir, S., and Sefa, I. (2016). Lyapunov-function and proportional-resonant-based control strategy for single-phase grid-connected VSI with LCL filter. *IEEE Trans. Ind. Electron.* 63 (5): 2838–2849.

80 Huang, M., Wang, X., Loh, P.C., and Blaabjerg, F. (2016). Active damping of LLCL-filter resonance based on LC-trap voltage or current feedback. *IEEE Trans. Power Electron.* 31 (3): 2337–2346.

81 Lei, Y., Xu, W., Mu, C. et al. (2014). New hybrid damping strategy for grid-connected photovoltaic inverter with LCL filter. *IEEE Trans. Appl. Superconduct.* 24 (5).

82 Mukherjee, N. and De, D. (2013). Analysis and improvement of performance in LCL filter-based PWM rectifier/inverter application using hybrid damping approach. *IET Power Electron.* 6 (2): 309–325.

83 Xu, J., Xie, S., and Tang, T. (2014). Improved control strategy with grid-voltage feedforward for LCL-filter-based inverter connected to weak grid. *IET Power Electron.* 7 (10): 2660–2671.

84 Li, W., Ruan, X., Pan, D., and Wang, X. (2013). Full-feedforward schemes of grid voltages for a three-phase LCL-type grid-connected inverter. *IEEE Trans. Ind. Electron.* 60 (6): 2237–2250.

85 Yang, D., Ruan, X., and Wu, H. (2014). Impedance shaping of the grid-connected inverter with LCL filter to improve its adaptability to the weak grid condition. *IEEE Trans. Power Electron.* 29 (11): 5795–5805.

86 Wang, X., Blaabjerg, F., and Loh, P.C. (2015). Virtual RC DAMPING of LCL-filtered voltage source converters with extended selective harmonic compensation. *IEEE Trans. Power Electron.* 30 (9): 4726–4737.

87 Shen, G., Zhu, X., Zhang, J., and Xu, D. (2010). A new feedback method for PR current control of LCL-filter-based grid-connected inverter. *IEEE Trans. Ind. Electron.* 57 (6): 2033–2041.

88 Timbus, A., Liserre, M., Teodorescu, R. et al. (2009). Evaluation of current controller for distributed power generation systems. *IEEE Trans. Power Electron.* 24 (3): 654–664.

89 Wu, W., Liu, Y., He, Y. et al. (2017). Damping methods for resonances caused by LCL-filter-based current-controlled grid-tied power inverters: an overview. *IEEE Trans. Ind. Electron.* 64 (9): 7402–7413.

90 Li, H., Wu, W., Huang, M. et al. (2020). Design of PWM-SMC controller using linearized model for grid-connected inverter with LCL filter. *IEEE Trans. Power Electron.* 35 (12): 12773–12786.

91 Zhao, J., Wu, W., Shuai, Z. et al. (2020). Robust control parameters design of PBC controller for LCL-filtered grid-tied inverter. *IEEE Trans. Power Electron.* 35 (8): 8102–8115.

92 Chen, X., Wu, W., Gao, N. et al. (2020). Finite control set model predictive control for LCL-filtered grid-tied inverter with minimum sensors. *IEEE Trans. Ind. Electron.* 67 (12): 9980–9990.

93 Luo, S., Wu, W., Koutroulis, E. et al. (2021). A new virtual oscillator control without third-harmonics injection for DC/AC inverter. *IEEE Trans. Power Electron.* 36 (9): 10879–10888.

2

Control Structure and Modulation Techniques of Single-Phase Grid-Connected Inverter

2.1 Control Structure of Single-Phase Grid-Connected Inverter

The application of renewable energy systems in power electronics is extensively studied to convert DC energy into the AC grid, where the control strategy of an inverter is directly related to the operational performance of the system. Figure 2.1 shows a general control block diagram of the grid-connected inverter with an *LCL*/*LLCL* filter in a distributed power generation system. The control of an *LCL*/*LLCL* filter filter-based inverter can be divided into grid-side current control and inverter-side current control.

Figure 2.1a shows the grid-side current control. The advantages of this control technique are fast dynamic response, high steady-state accuracy, and strong parameter robustness. However, if there is a control or sampling delay in the current control loop, the stability of the system will be affected.

Figure 2.1b shows the inverter-side current control by varying the amplitude and phase of the inverter output current to regulate the grid current. For indirect current control, it is advantageous to control system stability based on delay time [1]. However, the grid current feedback is not included in the control loop, and the resonance may be excited between the grid and filter capacitance resulting in a distorted injection current.

Single-phase inverters are classified into two types, namely single-stage and double-stage inverters. For double-stage inverter, MPPT controls the boost ratio of the DC/DC converter and DC-link voltage is controlled as an outer loop by the inverter. In previous studies, the control of grid-side controller is based on a DC-link voltage loop cascaded with an inner power loop instead of a current loop. Thus, the current injected into the utility network was indirectly controlled [2]. Moreover, control strategies employing an outer power loop and inner current

Control and Filter Design of Single-Phase Grid-Connected Converters, First Edition. Weimin Wu, Frede Blaabjerg, Henry Chung, Yuanbin He, and Min Huang.

Published 2023 by John Wiley & Sons, Inc.

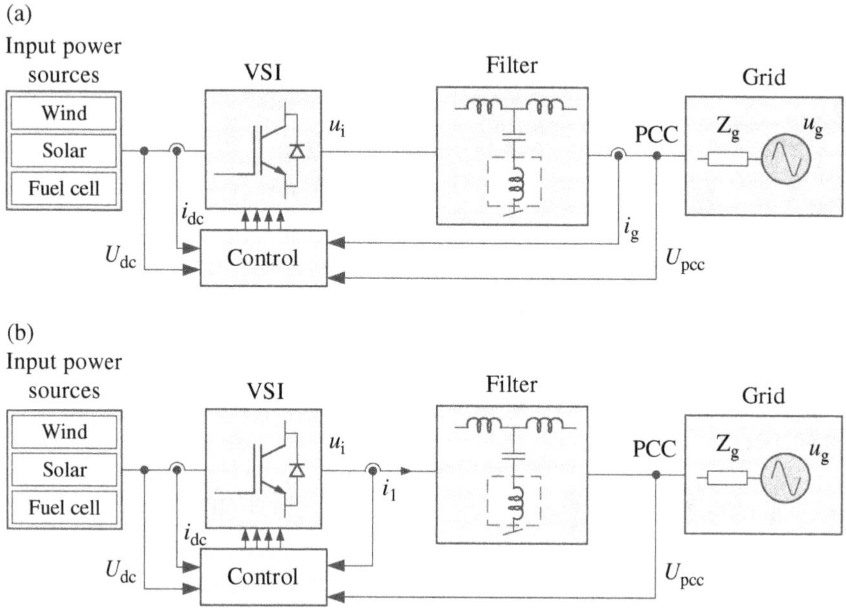

Figure 2.1 Schematic control diagram of single-phase grid-connected inverter with *LCL*/*LLCL* filter (a) grid-side current control; (b) inverter-side current control.

loop have also been reported [3]. If the DC voltage is high enough, MPPT directly controls the AC current as a single-stage inverter.

In the following sections, taken grid-side current control as an example, we discuss the classification of control strategies relative to the frame of reference in which they are implemented, and highlight important properties of each structure.

2.1.1 Natural Frame Control

Figure 2.2a shows the control diagram of the double-stage inverter in natural frame. The control strategy applied to this grid-connected inverter consists of two cascaded loops, namely fast internal current loop that regulates the grid current and an external voltage loop that controls the DC-link voltage [4, 5]. The current loop is responsible for resolving power quality issues and current protection; thus, harmonic compensation and dynamics are important properties of current controller. The DC-link voltage controller is designed to balance the power flow in the system. The output of DC-link voltage controller sets the active current reference. Using the phase angle of the grid voltages provided by a PLL system, the current reference is created and compared with the corresponding measured current resulting in an error in the controller.

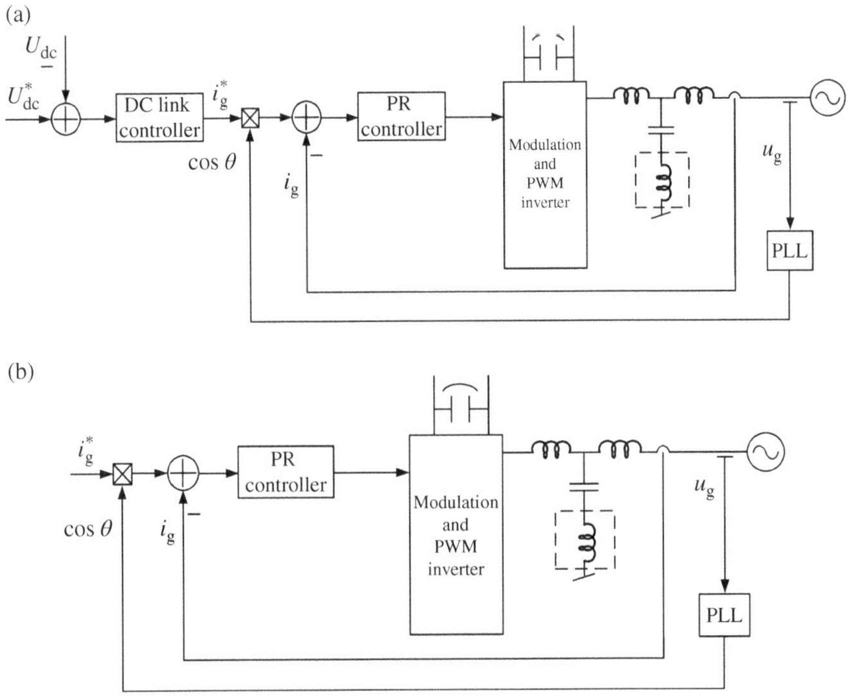

Figure 2.2 General structure of natural frame control structure for (a) double-stage inverter; (b) single-stage inverter.

Figure 2.2b shows the control diagram of single-stage inverter in natural frame. The current reference is directly compared with the corresponding measured current, and the phase angle of the grid voltages is provided by the PLL system.

Because the control variables are sinusoidal in this situation and the proportional–integral (PI) controller is not capable of removing the steady-state error when controlling sinusoidal waveforms, it is necessary to employ other types of controllers. One such controller is the proportional resonant (PR) controller [6, 7] that has gained popularity in the last decade in the current regulation of grid-connected systems. The controller matrix in the stationary reference frame of a PR controller is given by:

$$G_{PR}^{(\alpha\beta)}(s) = \begin{bmatrix} K_p + \dfrac{K_i s}{s^2 + \omega^2} & 0 \\ 0 & K_p + \dfrac{K_i s}{s^2 + \omega^2} \end{bmatrix} \tag{2.1}$$

where ω is the resonance frequency of the controller, K_p is the proportional gain, and K_i is the integral gain of the controller. A characteristic of this controller is that it achieves a very high gain around the resonance frequency, thus eliminating the steady-state error between the controlled signal and its reference. The width of the frequency band around the resonance point depends on the integral time constant K_i. A low K_i results in a narrow band whereas high K_i results in a wider band. Moreover, the high dynamic characteristics of PR controllers have been reported in previous studies [6–8].

2.1.2 Synchronous Reference Frame Control

Figure 2.3a shows the control diagram of double-stage inverter in synchronous reference frame. Synchronous reference frame control, also known as dq control, uses a reference frame transformation module to transform the grid current and voltage waveforms into a reference frame that rotates synchronously with the grid voltage. The control variables become dc values; thus, filtering and controlling can be easily achieved. In this structure, the DC-link voltage is controlled in accordance with the necessary output power or DC-link voltage reference. Its output is the reference for the active current controller whereas the reference for reactive current is usually set to zero if reactive power control is not allowed.

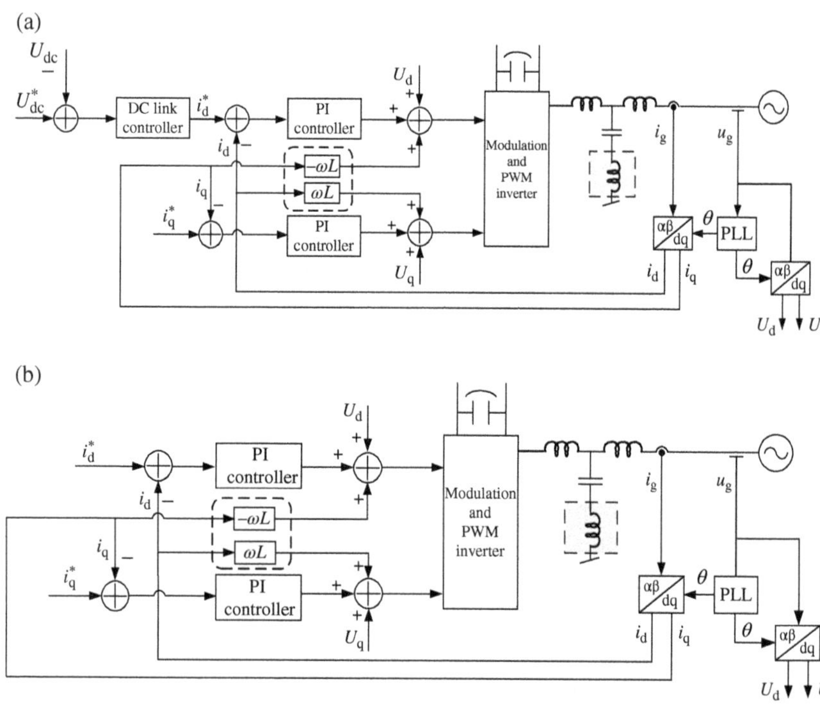

Figure 2.3 General structure of synchronous rotating frame control structure for (a) double-stage inverter; (b) single-stage inverter.

A schematic of single-stage inverter in synchronous reference frame control is shown in Figure 2.3b. Current references are directly provided to control AC current.

The dq control structure is normally associated with PI controllers because they exhibit a satisfactory behavior when DC variables are regulated. The matrix transfer function of the controller in dq coordinates can be written as:

$$G_{\mathrm{PI}}^{(\mathrm{dq})}(s) = \begin{bmatrix} K_{\mathrm{p}} + \dfrac{K_{\mathrm{i}}}{s} & 0 \\ 0 & K_{\mathrm{p}} + \dfrac{K_{\mathrm{i}}}{s} \end{bmatrix} \tag{2.2}$$

where K_{p} is the proportional gain and K_{i} is the integral gain of the controller. Because the controlled current must be in phase with the grid voltage, the phase angle used by the $\alpha\beta \rightarrow \mathrm{dq}$ transformation module must be extracted from the grid voltages. Hence, filtering the grid voltages and using the arctangent function to extract the phase angle can be a possibility [9]. In addition, the phase-locked loop (PLL) technique is a state-of-the-art technique for extracting the phase angle of the grid voltages in distributed generation systems.

To improve the performance of PI controller in the control structure depicted in Figure 2.3, cross-coupling terms and voltage feedforward are usually used [5, 10]. However, with all these improvements, the compensation capability of the low-order harmonics in PI controllers is very poor, which is a major drawback when they are used in grid-connected systems.

2.1.3 Grid Synchronization Methods

According to the standards of field request, the current injected into the utility network must be synchronized with the grid voltage [11, 12]. Therefore, grid synchronization algorithms play an important role in distributed-generation systems. The output of the synchronization algorithm is the phase of the grid voltage vector, which is used to synchronize the control variables; for example, grid currents with grid voltages using different transformation modules such as $\alpha\beta \rightarrow \mathrm{dq}$.

To date, many previous studies have developed different methods to extract phase angles. In [13], a comparison of the techniques used for detecting the phase angle of grid voltages under different grid conditions was performed. Advantages and disadvantages as well as an evaluation of the performance are also presented. This section provides a brief description of these methods.

2.1.3.1 Zero-Crossing Method

Among all the techniques, the zero-crossing method has the simplest implementation. However, poor performance is also reported if grid voltages register variations, such as harmonics or notches.

2.1.3.2 Filtering of Grid Voltages

Even though this method performs better than the zero-crossing method, it is difficult to extract the phase angle when grid variations or faults occur in the utility network [3]. Hence, arctangent function is used to obtain the phase angle of the utility voltage. However, filtering introduces a delay in the processed signal, which is unacceptable when it is used for extracting the grid voltage angle. Therefore, a proper filter design is necessary.

When the current controller is implemented in the stationary reference frame, knowledge of the grid voltage angle θ is not required; hence, it is not necessary to calculate the arctangent function. In fact, the filtered αβ components can be directly used as a template for synchronizing the reference current signal.

2.1.3.3 PLL Technique

Currently, the PLL technique is the state-of-the-art method for extracting the phase angle of the grid voltages [3]. It is implemented in a dq synchronous reference frame and its schematic is illustrated in Figure 2.4.

The phase detector (PD) generates an output signal proportional to the phase difference between v and v'. Depending on the type of PD, high-frequency AC components appear along with the DC phase-difference signal. The loop filter (LF) exhibits low-pass characteristics and filters out the high-frequency AC components from the PD output. Typically, this is a first-order low-pass filter or PI controller. A voltage-controlled oscillator (VCO) generates an ac signal at its output whose frequency varies with respect to the central frequency as a function of the input voltage.

As shown in Figure 2.3, the PLL is realized in a synchronous rotating frame for a single-phase inverter. A regulator, usually PI, is used to control this variable and the output of this regulator is the grid frequency. Then, the grid frequency is integrated to obtain the utility voltage angle that is fed back into the αβ → dq transformation module to transform into the synchronous rotating reference frame. This algorithm has a better rejection of grid harmonics, notches, and any other kind of disturbance; however, additional improvements must be made to

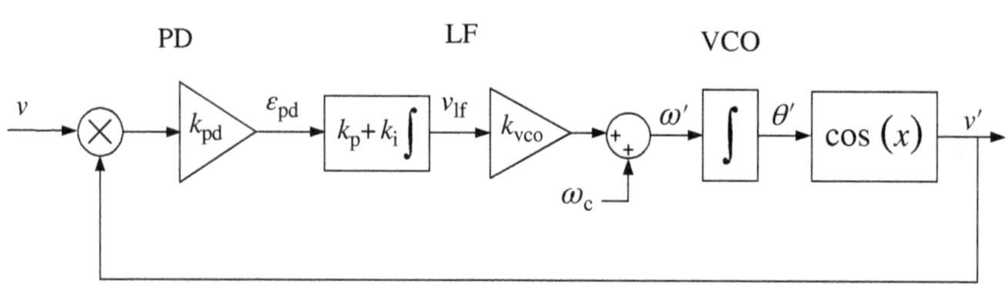

Figure 2.4 Basic PLL block.

overcome grid imbalance. In the case of unsymmetrical voltage faults, the second harmonics produced by the negative sequence will propagate through the PLL system and reflected in the extracted phase angle. To overcome this, different filtering techniques are necessary such that the negative sequence is filtered out. Consequently, during unbalanced conditions, the three-phase dq PLL structure can estimate the phase angle of the positive sequence of the grid voltages.

2.2 Modulation Methods

Figure 2.5 shows the topology of a single-phase grid-connected inverter with an *LCL*/*LLCL* filter. For the convenience of analysis, the left and right bridge arms of the full-bridge inverter are divided into two bridge arms A and B. According to the output voltage level of the inverter, the modulations of a single-phase inverter are classified into two categories: three-level modulation and two-level modulation, which are also called unipolar modulation and bipolar modulation. Figure 2.6 shows the equivalent circuit diagram of a single-phase grid-connected inverter with an *LCL*/*LLCL* filter. u_i is the output of the inverter after SPWM and u_g is the grid voltage.

2.2.1 Unipolar Modulation Method

Based on the switching sequence during the fundamental cycle, unipolar modulation can also be divided into two types: continuous and discontinuous unipolar modulation [14].

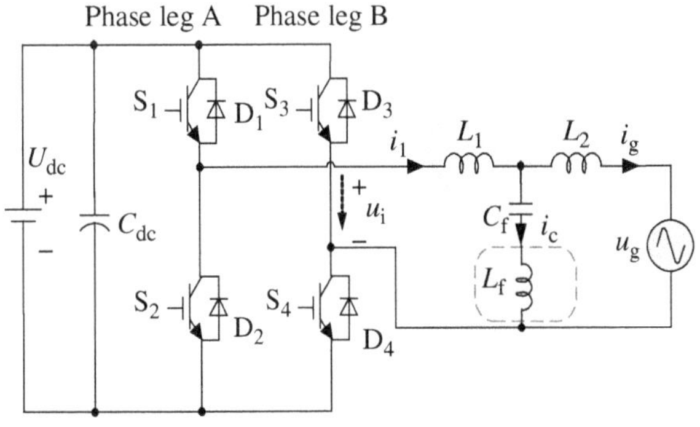

Figure 2.5 Topology of single-phase grid-connected inverter with *LCL*/*LLCL* filter.

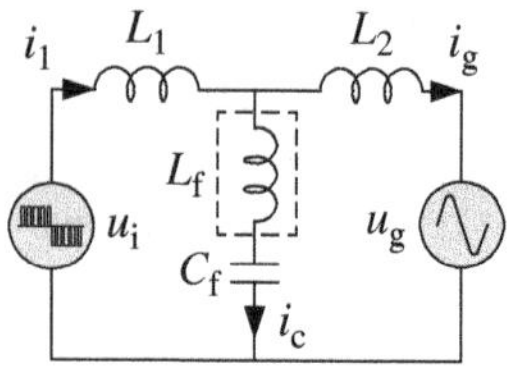

Figure 2.6 Equivalent circuit diagram of single-phase grid-connected inverter with *LCL*/*LLCL* filter.

2.2.1.1 Continuous Unipolar Modulation

For continuous unipolar modulation, each phase leg of the inverter switches between the upper and lower DC rails continuously over the fundamental cycle as the carrier waveform ramps above and below the reference waveform. Figure 2.7 shows the reference waveform and the carrier and driving signals of the inverter under continuous unipolar modulation where v_{s1}, v_{s2}, v_{s3}, and v_{s4} are the driving signals for the corresponding switches as shown in Figure 2.7. The driving signals of the same phase leg are complementary. The switching frequency of current in the inverter-side inductor is twice that of the carrier frequency owing to the interaction between the phase-leg transitions. However, the size and volume of filter inductors and capacitors can be reduced.

2.2.1.2 Discontinuous Unipolar Modulation

Discontinuous unipolar modulation is divided into two categories as shown in Figure 2.8. Figure 2.8a depicts a cost-effective approach where only phase leg B is modulated and phase leg A switches between the negative and positive DC rail for each half of the fundamental cycle. Figure 2.8b depicts an approach in

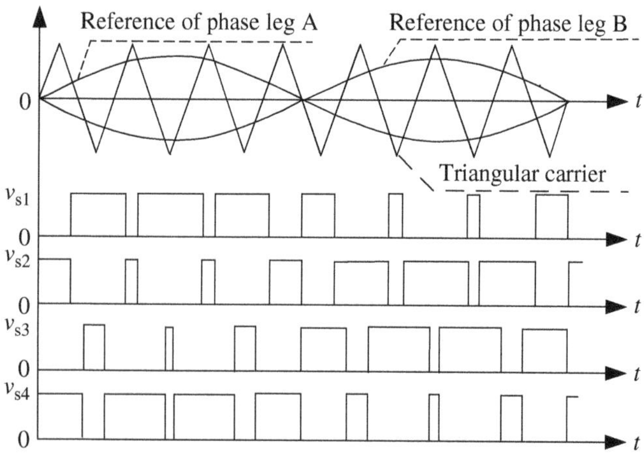

Figure 2.7 Voltage reference, carrier, and driving signals under continuous unipolar modulation.

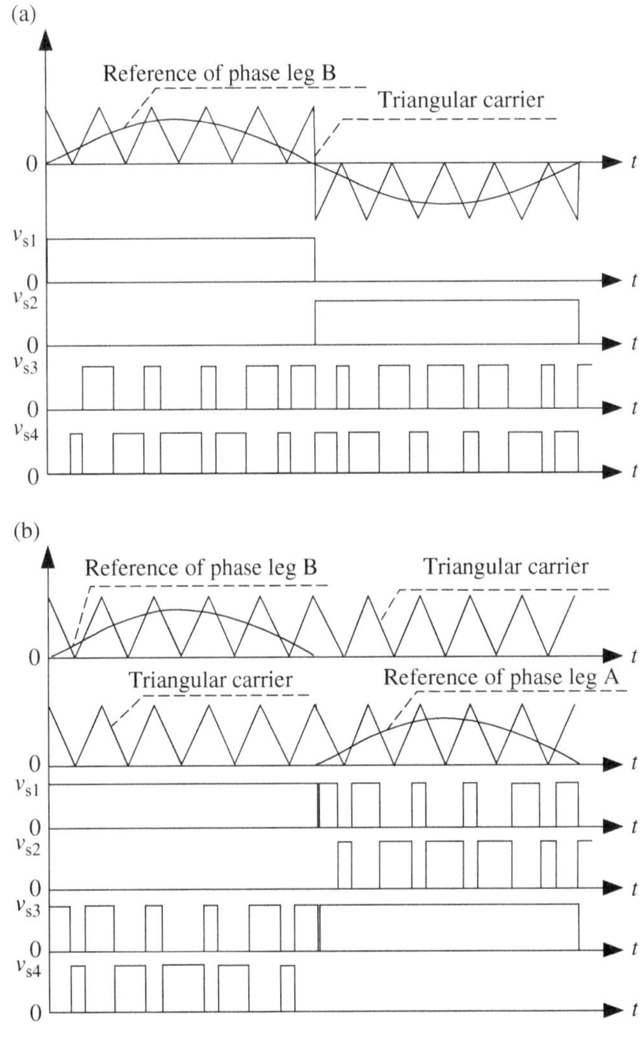

Figure 2.8 Voltage reference, carrier, and driving signals under discontinuous unipolar modulation: (a) single phase leg B is modulated; (b) both phase legs are modulated.

which both phase legs must be modulated and demands a certain level of complexity for the controller implementation. In this approach, only phase leg B is modulated in the first half of the fundamental cycle and phase leg A is modulated during the second half of the fundamental cycle. The phase legs never switch simultaneously within a single carrier period under this modulation. Compared to continuous modulation, the output pulses of the inverter occur only at the carrier frequency because, only one phase leg chops at high frequency during one half of the fundamental cycle.

The output voltage u_i of the inverter can be decomposed according to the double Fourier transform with discontinuous unipolarity as follows [14]:

$$u_i(t) = \alpha U_{dc} \cos(\omega_o t) + \sum_{m=1}^{\infty} \sum_{n=\pm 1}^{\pm\infty} \frac{2U_{dc} J_n(m\pi\alpha)}{m\pi} \sin\left(\frac{n\pi}{2}\right) \cos(m\omega_s t + n\omega_o t) \tag{2.3}$$

where α is the modulation index, U_{dc} is the DC-link voltage, ω_o is the fundamental frequency in radians per second, ω_s is the switching frequency in radians per second and $J_n(x)$ is the integral of the Bessel function depicting different sideband harmonic magnitudes.

According to function (2.3), Figure 2.9 depicts the spectrum of the inverter harmonic voltage for discontinuous unipolar SPWM modulation. The modulation index of the inverter is 0.899, fundamental wave frequency is 50 Hz, switching frequency is 20 kHz, and DC bus voltage is 350 V. The harmonic spectrum shows that for a single-phase inverter, the dominant harmonics with discontinuous unipolar SPWM are distributed around the carrier frequency.

There is another modulation method for a single-phase voltage source inverter called dual-buck modulation. It is similar to discontinuous unipolar modulation; however, the difference is that only one switch of each phase leg is modulated and

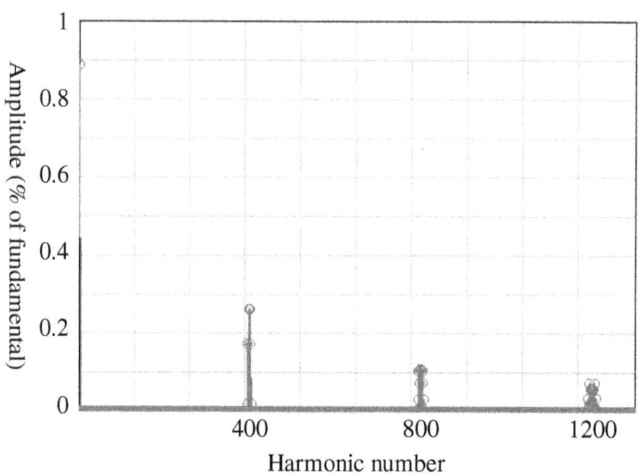

Figure 2.9 Spectrum of inverter harmonic voltage for unipolar SPWM modulation.

the other switch is off. In this case, the inverter works like a pure buck converter. Therefore, the current through the inverter-side inductor cannot be reversed during each carrier period.

2.2.2 Bipolar Modulation Method

The output voltage of single-phase inverter produced by bipolar modulation is quite different from unipolar modulation because it continuously switches between positive and negative active pulses in each carrier period without any zero states. This is known as two-level modulation. As shown in Figure 2.10, bipolar modulation can be implemented with very simple circuitry for one phase leg to maintain the inverse switch state of the other. For a single-phase inverter, the dominant harmonics with bipolar SPWM are distributed around the carrier frequency.

The output voltage u_i of the inverter with natural sampling (SPWM) by a bipolar triangular wave and sinusoidal wave is given by:

$$u_i(t) = \alpha U_{dc}\cos(\omega_0 t + \theta_0) + \sum_{m=1}^{\infty} \frac{4U_{dc}J_0\left(m\frac{\pi}{2}a\right)}{m\pi}\sin\left(m\frac{\pi}{2}\right)\cos\left[m(\omega_s t + \theta_c)\right]$$
$$+ \sum_{m=1}^{\infty}\sum_{n=\pm 1}^{\pm\infty} \frac{4U_{dc}J_n\left(n, m\frac{\pi}{2}\alpha\right)}{m\pi}\sin\left[(m+n)\frac{\pi}{2}\right]\cos\left[m(\omega_s t + \theta_c) + n(\omega_0 t + \theta_0)\right] \tag{2.4}$$

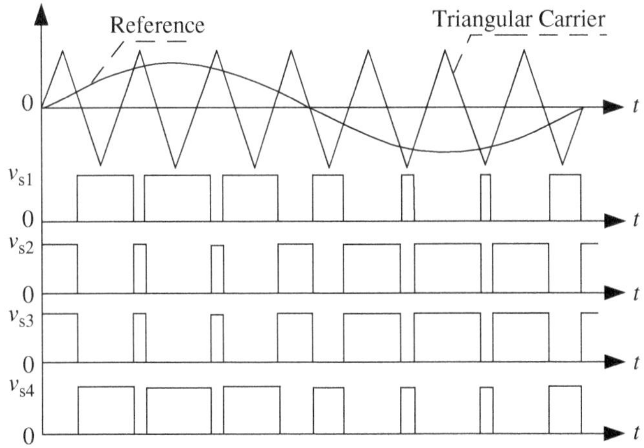

Figure 2.10 Voltage reference, carrier, and driving signals under bipolar modulation.

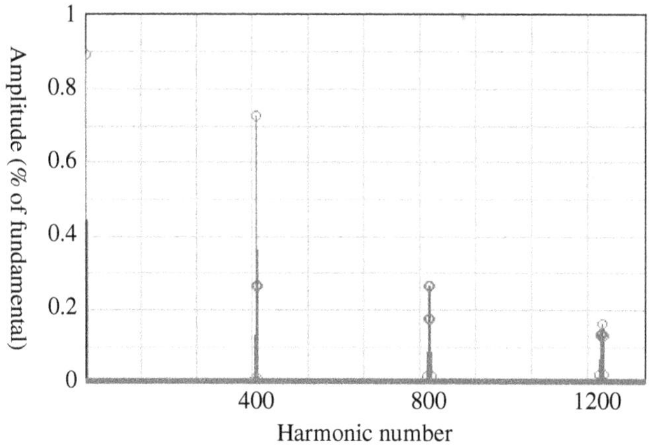

Figure 2.11 Spectrum of inverter harmonic voltage for bipolar SPWM modulation.

Figure 2.11 shows the spectrum of the inverter harmonic voltage for bipolar SPWM modulation based on function (2.4). The harmonic spectrum shows that for a single-phase inverter, the bipolar SPWM generates higher harmonics than the unipolar method. Hence, a larger inductance is required to achieve the same current ripple for bipolar SPWM modulation.

2.3 Summary

In this chapter, the control structure and modulation techniques of a single-phase grid-connected inverter were presented. Different implementation structures, such as dq and natural frame control structures for single-stage and double-stage inverters, were analyzed and their major characteristics were identified. A brief overview of the grid synchronization algorithms was also provided. For the double-stage inverter, it was observed that the boost ratio of the DC/DC converter was controlled by MPPT and DC-link voltage was controlled as an outer loop by the inverter. However, if the DC voltage was very large or varied significantly, MPPT directly controlled the AC current as a single-stage inverter. The research presented in this book primarily focuses on single-stage inverters.

Additionally, modulation techniques of single-phase grid-connected inverters based on the output voltage level of the inverter were discussed. They are classified as unipolar and bipolar modulation. Different modulation techniques result in different design parameters of the *LCL*/*LLCL* filter. It was observed that bipolar modulation requires a larger inductance than unipolar modulation to achieve the same harmonic attenuation; however, the modulation technique is simple, which is beneficial for the suppression of common-mode interference. In the next chapter, the filter design procedure has been discussed in detail.

References

1 Zou, C., Liu, B., Duan, S., and Li, R. (2014). Influence of delay on system stability and delay optimization of grid-connected inverters with LCL filter. *IEEE Trans. Ind. Informat.* 10 (3): 1775–1784.

2 Ramos, C., Martins, A., and Carvalho, A. (2002). Current control in the grid connection of the double-output induction generator linked to a variable speed wind turbine. *Proceedings of the IEEE IECON*, vol. 2, pp. 979–984, Seville, Spain (5–8 November 2002). New York: IEEE.

3 Blaabjerg, F., Teodorescu, T., Liserre, M., and Timbus, A.V. (2006). Overview of control and grid synchronization for distributed power generation systems. *IEEE Trans. Ind. Electron.* 53 (5): 1398–1409.

4 Agirman, I. and Blasko, V. (2003). A novel control method of a VSC without ac line voltage sensors. *IEEE Trans. Ind. Appl.* 39 (2): 519–524.

5 Teodorescu, R. and Blaabjerg, F. (2004). Flexible control of small wind turbines with grid failure detection operating in stand-alone or grid-connected mode. *IEEE Trans. Power Electron.* 19 (5): 1323–1332.

6 Wang, X., Blaabjerg, X.F., and Loh, P.C. (2016). Grid-current-feedback active damping for LCL resonance in grid-connected voltage source inverters. *IEEE Trans. Power Electron.* 3 (1): 213–223.

7 Komurcugil, H., Altin, N., Ozdemir, S., and Sefa, I. (2016). Lyapunov-function and proportional-resonant-based control strategy for single-phase gridconnected VSI with LCL filter. *IEEE Trans. Ind. Electron.* 63 (5): 2838–2849.

8 Zmood, D. and Holmes, D.G. (2003). Stationary frame current regulation of PWM inverters with zero steady-state error. *IEEE Trans. Power Electron.* 18 (3): 814–822.

9 Gardner, F.M. (1979). *Phase Lock Techniques*. New York: Wiley.

10 Liserre, M., Teodorescu, R., and Blaabjerg, F. (2006). Stability of photovoltaic and wind turbine grid-connected inverters for a large set of grid impedance values. *IEEE Trans. Power Electron.* 21 (1): 263–272.

11 Characteristic of the Utility Interface for Photovoltaic (PV) Systems IEC 61727 CDV (Committee Draft *for* Vote). 2002.

12 IEEE Std 1547TM-2003 (2005). *IEEE Standard for Interconnecting Distributed Resources with Electric Power Systems*. New York: IEEE.

13 Timbus, A.V., Liserre, M., Teodorescu, R., and Blaabjerg, F. (2005). Synchronization methods for three phase distributed power generation systems. An overview and evaluation. *Proceedings of the IEEE PESC*, pp. 2474–2481, Dresden, Germany (16 June 2005). New York: IEEE.

14 Holmes, D.G. and Lipo, T.A. (2003). *Pulse Width Modulation for Power Converters: Principles and Practice*. Hoboken, NJ: Wiley.

Part II

LCL/LLCL Power Filter

3

An *LLCL* Power Filter for Single-Phase Grid-Connected Inverter

3.1 Introduction

With an increasing attention to renewable energy power generation, grid-connected inverters are widely used [1–3]. Low-pass power filters are usually inserted between the voltage source inverter (VSI) and grid to limit excessive current harmonics that are caused by sinusoidal pulse width modulation and injected into the connection point.

Owing to the rise in copper price, researchers have taken many measures to reduce the cost of power filters. An effective method is to increase the switching frequency of the inverter. However, this solution depends on the equipment technology and cost. For example, compared to silicon devices with identical power ratio, the switching frequency of silicon carbide devices is much higher; however, the price is much higher. Another measure focuses on special topologies or controls such as dual-mode time-sharing inverters [4] to utilize the good performance of low-voltage equipment to achieve higher switching frequencies. The dual-mode time-sharing control method of single-phase [5] and three-phase [6] inverters maximizes the modulation index and reduces the size of the output power filter; however, the topology or control becomes more complicated resulting in reliability drops. In addition, it is difficult for dual-mode time-sharing inverters to compensate for harmonics or generate reactive power for the grid. The most common solution is to use a third-order LCL filter to replace the first-order *L* filter [7]. *LCL* filters can meet grid-connected standards with a smaller size and cost especially in applications above several kilowatts, but it may be more difficult to maintain system stability [8–10]. Moreover, compared to the *L* filter, the parameter selection of *LCL* filter [11, 12] is also a complicated process.

This chapter proposes a novel topology of high-order power filters for grid-connected voltage source inverters known as *LLCL* filter. A small inductor is inserted in the branch loop of the capacitor of the traditional *LCL* filter to form

Control and Filter Design of Single-Phase Grid-Connected Converters, First Edition. Weimin Wu, Frede Blaabjerg, Henry Chung, Yuanbin He, and Min Huang.

Published 2023 by John Wiley & Sons, Inc.

a series resonant circuit at the switching frequency. In particular, it can attenuate the current ripple component near the switching frequency better than the *LCL* filter thereby saving the total inductance and reducing the size.

The rest of the chapter is organized as follows: Section 3.2 describes the principle of traditional *LCL* filter and proposed *LLCL* filter. The parametric design of *LLCL* filters is introduced in Section 3.3. It is divided into the following three aspects: constraints and procedure on the power filter design; saving analysis on the grid-side inductance and specific design consideration for a simple passive damping strategy. Section 3.4 describes the design examples of *LCL* and *LLCL* filters. Sections 3.5 and 3.6 present the experimental results and summary, respectively.

3.2 Principle of Traditional *LCL* Filter and Proposed *LLCL* Filter

In Chapter 2, single-phase grid-connected VSI based on *LCL* filter has been discussed. The equivalent circuit of an *LCL*-filter-based single-phase grid-connected VSI is shown in Figure 3.1. The inverter output voltage and current are represented as u_i and i_1, respectively, and the grid voltage and current are represented as u_g an i_g, respectively. The switching frequency is represented as f_s (in hertz) or ω_s (in radians per second). It is assumed that the power grid is an ideal voltage source (i.e., zero impedance) to supply a constant voltage at the fundamental frequency of 50 or 60 Hz.

When the single-phase full-bridge VSI is operated under the condition of unipolar, sine-triangle, and asymmetrical regular-sampled PWM, its output voltage $u_i(t)$ can be derived as: [13]

$$u_i(t) = \alpha U_{dc} \cos(\omega_o t) + \sum_{k=1}^{\infty} \sum_{n=\pm 1}^{\pm\infty} \frac{2U_{dc} J_n(k\pi\alpha)}{k\pi} \sin\left(\frac{n\pi}{2}\right) \cos(k\omega_s t + n\omega_o t) \tag{3.1}$$

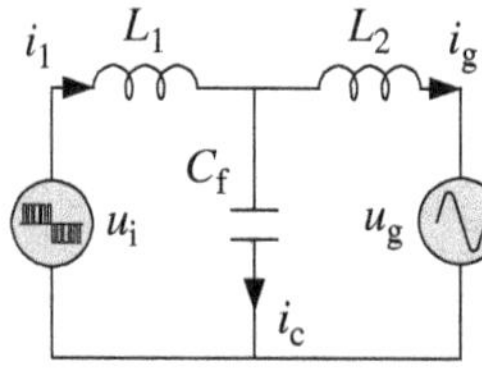

Figure 3.1 Equivalent circuit of an *LCL*-filter-based single-phase grid-connected VSI.

where α is the modulation index, U_{dc} is the DC-link voltage, ω_o is the fundamental frequency in radians per second, and $J_n(x)$ is the integral of the Bessel function, which is expressed as $J_n(x) = \frac{1}{\pi}\int_0^{\pi} \cos(n\pi - x\sin\tau)d\tau$ highlighting the different side-band harmonic magnitude.

The inverter output impedance can be described as:

$$Z_o(j\omega) = \left.\frac{u_i(j\omega)}{i_1(j\omega)}\right|_{\substack{u_g(j\omega)=0 \\ \omega \neq \omega_o}} = \frac{L_1 L_2 C_f (j\omega)^3 + (L_1 + L_2) j\omega}{L_2 C_f (j\omega)^2 + 1} \tag{3.2}$$

Based on the harmonic current recommendation in IEEE 1547.2-2008 and 519-1992 [14, 15], it is reasonable to consider the grid-side current i_g as the "ideal" constant current only at the fundamental frequency. In other words, the branch circuit of inductor L_2 is open while considering the effects of the inverter high-frequency harmonics. Then, (3.2) can be rewritten as:

$$Z_o(j\omega)\big|_{\omega\neq\omega_o} \cong L_1 j\omega + \frac{1}{C_f j\omega} \tag{3.3}$$

Furthermore, in an *LCL* filter, $\frac{1}{C_f j\omega} << L_1 j\omega$ is generally true around the switching frequency or higher. Then, (3.3) can be simplified as:

$$Z_o(j\omega)\big|_{\omega\neq\omega_o} \cong L_1 j\omega \tag{3.4}$$

The harmonic amplitudes of inverter-side current i_1 can be derived as:

$$|I_{AM}|\big|_{\omega\neq\omega_o} = \frac{|U_{AM}(n,\ k)|}{|Z_o(j\omega)|} \tag{3.5}$$

where the amplitude of inverter output harmonic voltage $U_{AM}(n, k)$is given by:

$$U_{AM}(n,\ k)\big|_{\substack{k=1,2,\cdots,\infty \\ n=\pm1,\pm2,\cdots,\pm\infty}} = \left|\frac{2U_{dc} J_n(k\pi\alpha)}{k\pi} \sin\left(\frac{n\pi}{2}\right)\right| \tag{3.6}$$

Figure 3.2 shows the main harmonic current spectrum of the inverter output current under the condition that the modulation index α is 0.9, DC-link voltage U_{dc} is 350 V, max inverter-side current ripple is 31.5% I_{ref} (where I_{ref} is the reference fundamental peak current), and switching frequency f_s is 20 kHz. It can be seen that the uppermost harmonics of inverter output current are around the switching frequency. Therefore, the grid-side inductor L_2 and paralleled capacitor C_f are both limited by the harmonics around the switching frequency [11, 12, 16, 17].

We have assumed that the zero impedance in the branch circuit of the capacitor appears at the switching frequency. Then, the grid-side inductor L_2 would be sized based on twice the switching-frequency current attenuation from the inverter side to the grid side, which is very beneficial to scale down the grid-side inductance. To achieve this, a novel higher-order power filter has been proposed in the next section.

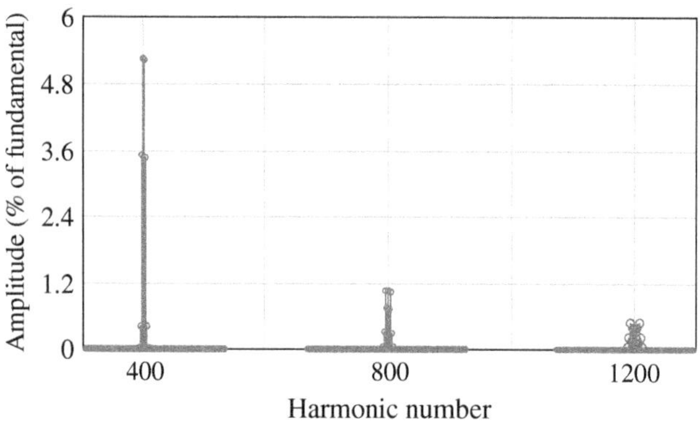

Figure 3.2 Main harmonic current spectrum of inverter with unipolar modulation.

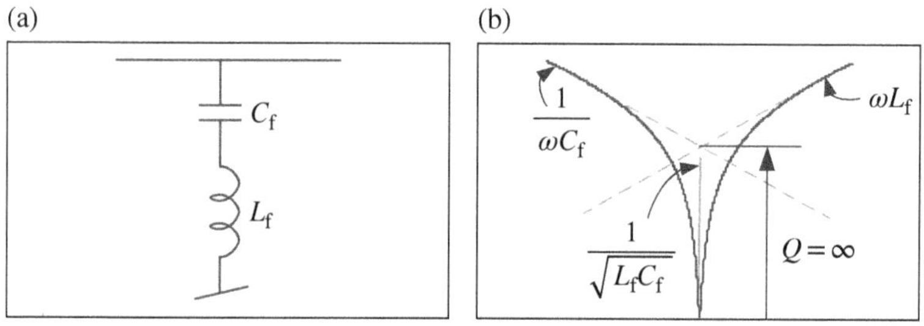

Figure 3.3 (a) Second-order series resonant filter; (b) second-order filter impedance transfer functions.

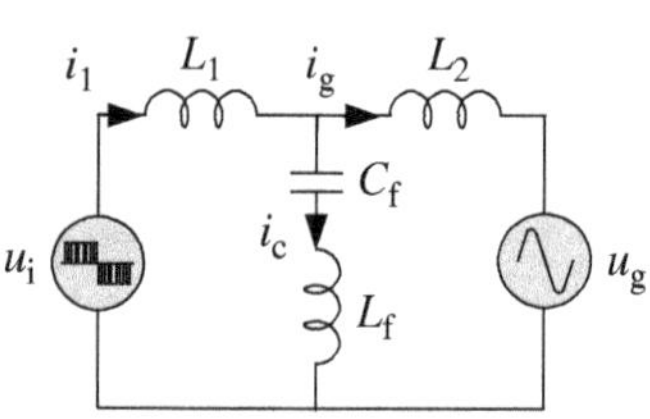

Figure 3.4 Schematic diagram of the *LLCL* filter.

To achieve zero impedance of the capacitor branch circuit at the switching frequency, the ideal series resonance shown in Figure 3.3 performs efficiently. Hence, a novel high-order power filter known as *LLCL* filter is proposed as shown in Figure 3.4. Compared to the *LCL* filter, an *LLCL* filter consists of a small inductor that is inserted in the branch loop of the capacitor composed of an $L_f - C_f$ series resonant circuit at the switching frequency, and its inductance is generally much smaller than that of L_1 or L_2.

Assuming that the grid is an ideal sinusoidal voltage source, the transfer functions $\frac{i_1(s)}{u_i(s)}$ and $\frac{i_g(s)}{u_i(s)}$ of *LLCL* filter can be derived as:

$$G_{u_i \to i_{L_1}}(s) = \left.\frac{i_1(s)}{u_i(s)}\right|_{u_g(s)=0} = \frac{(L_2 + L_f)C_f s^2 + 1}{(L_1 L_2 C_f + (L_1 + L_2)L_f C_f)s^3 + (L_1 + L_2)s} \tag{3.7}$$

$$G_{u_i \to i_g}(s) = \left.\frac{i_g(s)}{u_i(s)}\right|_{u_g(s)=0} = \frac{L_f C_f s^2 + 1}{(L_1 L_2 C_f + (L_1 + L_2)L_f C_f)s^3 + (L_1 + L_2)s} \tag{3.8}$$

If the inductance of L_f is set to zero, then the transfer functions of *LCL* filter can be calculated.

Figure 3.5 shows the transfer functions $\frac{i_1(s)}{u_i(s)}$ and $\frac{i_g(s)}{u_i(s)}$ of both *LLCL* filter and *LCL* filter. All the other parameters are the same except for L_f. It can be seen that within half of the switching frequency range, an *LLCL*-filter-based grid-connected VSI has almost the same frequency-response characteristic as that of an *LCL* filter. That is, compared to the *LCL* filter, the additional inductor L_f of the *LLCL* filter does not bring any extra control difficulties. Furthermore, a smaller value of the grid-side inductor L_2 of the *LLCL* filter could be beneficial to widen the control bandwidth.

3.3 Parametric Design of *LCL* and *LLCL* Filters

3.3.1 Constraints and Procedure of Power Filter Design

When designing the power filters (whether it is an *LCL* or *LLCL* filter), certain limitations must be introduced on the parameter values as discussed in [12, 16].

1) The capacitor value C_f is limited to a value less than 5% to ensure that the capacitive reactive power is not exceeded at rated load.
2) The upper limit of the total inductance $(L_1 + L_2)$ depends on the voltage drop during operation (lower than 10%).
3) The value of the inverter-side inductor L_1 is limited to the requirement of the ripple current (generally lower than 40%).
4) IEEE 519-1992 recommends that harmonics higher than the 35th should be limited. For a grid-connected inverter system, if the short-circuit current of power system is lower than 20 times the nominal grid-side fundamental

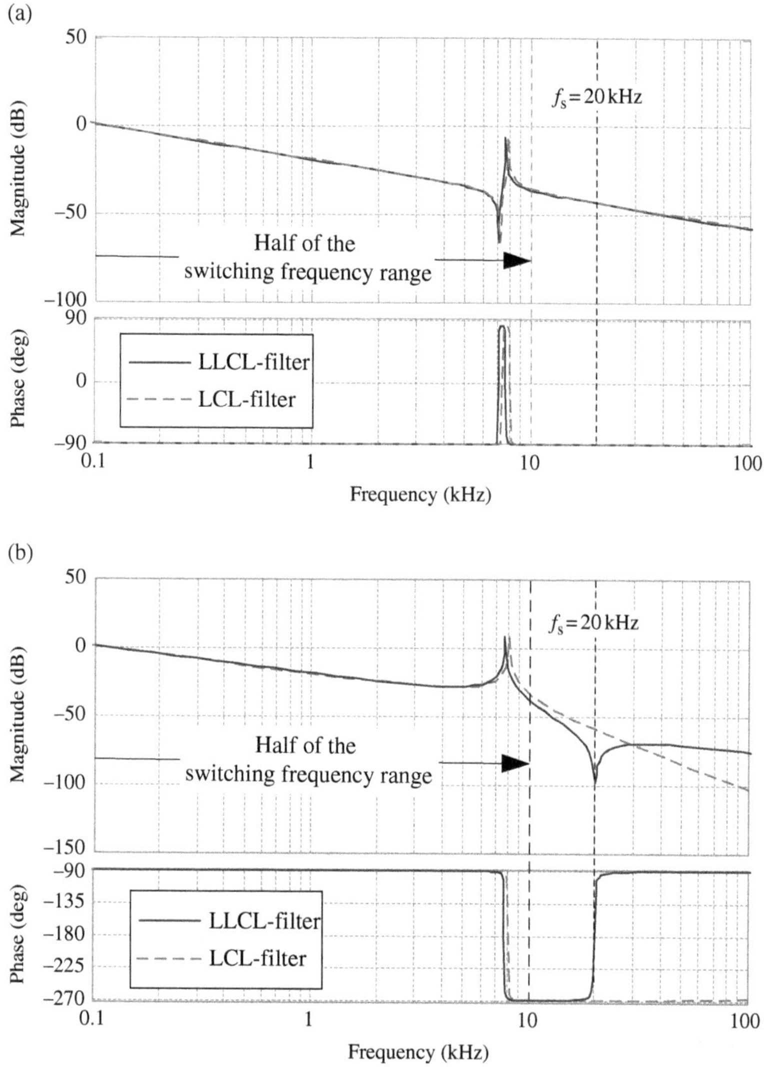

Figure 3.5 Bode plots of transfer functions in *LCL* and *LLCL* filters. (a) Bode plots of transfer functions $i_1(s)/u_i(s)$; (b) Bode plots of transfer functions $i_g(s)/u_i(s)$.

current, then each harmonic current higher than the 35th should be less than 0.3% of the rated fundamental current.

5) The resonance frequency f_r or ω_r should be in the range between ten times the line frequency and half of the switching frequency to avoid resonance problems in the lower and higher parts of the harmonic spectrum.

Based on the aforementioned desired limits, the filter can be designed with the following seven-step design procedure.

1) Choice of the tolerable current ripple on the inverter side → design of the inner inductor L_1 is given by:

$$15\% \leq \frac{\Delta I_1}{I_{\text{ref}}} = \frac{U_{\text{dc}}}{4L_1 f_s I_{\text{ref}}} \leq 40\% \tag{3.9}$$

where ΔI_1 is the inverter-side current ripple and I_{ref} is the rated reference peak current.

2) Choice of the reactive power absorbed under the rated condition → capacitor value. Considering x as the percentage of the reactive power absorbed at rated conditions, capacitor value is given by:

$$C_{\text{f}} = xC \tag{3.10}$$

where x is less than one. The upper capacitor value is limited by the condition given as:

$$C = \frac{5\% P_{\text{rated}}}{U_{\text{g}}^{2} \omega_{\text{o}}} \tag{3.11}$$

where P_{rated} is the rated power and U_{g} is the grid fundamental RMS voltage.

3) As far as an *LLCL* filter is concerned, the values of L_1 and C_{f} chosen are the same as in an *LCL* filter to meet the requirements of the inverter's current ripple and reactive power limitation, respectively. Since $L_{\text{f}} - C_{\text{f}}$ circuit resonates at the switching frequency, L_{f} can be considered as:

$$\frac{1}{\sqrt{L_{\text{f}} C_{\text{f}}}} = \omega_{\text{s}} \tag{3.12}$$

4) For an *LCL* filter, L_2 depends on the objective to attenuate each harmonic around the switching frequency down to 0.3%, which can be described using (3.13):

$$\frac{\frac{2U_{\text{dc}}}{\pi} \times \max\left(|J_1(\pi\alpha)|, |J_3(\pi\alpha)|\right) \times \left| G_{u_{\text{i}} \to i_{\text{g}}}(j\omega_{\text{s}}) \big|_{L_{\text{f}} = 0} \right|}{I_{\text{ref}}} \leq 0.3\% \tag{3.13}$$

where $J_1(\pi\alpha)$ and $J_3(\pi\alpha)$ are the integrals of the Bessel function corresponding to the sideband harmonics at the frequency of $(\omega_{\text{s}} + \omega_{\text{o}})$ and $(\omega_{\text{s}} + 3\omega_{\text{o}})$. For an *LLCL* filter, the current harmonics around the switching frequency satisfy the requirements of IEEE 519-1992 with far more ease owing to the $L_{\text{f}} - C_{\text{f}}$

resonant circuit, and the dominant ones appear around twice the switching frequency. Therefore, the important criteria for choosing L_2 of an *LLCL* filter should be expressed as in (3.14):

$$\frac{\frac{U_{dc}}{\pi} \times \max\left(|J_1(2\pi\alpha)|, |J_3(2\pi\alpha)|, |J_5(2\pi\alpha)|\right) \times \left|G_{u_i \to i_g}(j2\omega_s)\right|}{I_{ref}} \leq 0.3\% \tag{3.14}$$

where $J_1(2\pi\alpha)$, $J_3(2\pi\alpha)$, and $J_5(2\pi\alpha)$ are the integrals of the Bessel function corresponding to the sideband harmonics at the frequency of $(2\omega_s + \omega_o)$, $(2\omega_s + 3\omega_o)$, and $(2\omega_s + 5\omega_o)$. If the sum of the two inductances does not fulfill the requirement 2, a different value of attenuation of the absorbed reactive power should be selected and return to Step 2.

5) Verify the resonance frequency obtained.

$$\omega_r = \frac{1}{\sqrt{\left(\frac{L_1 L_2}{L_1 + L_2} + L_f\right) C_f}} \tag{3.15}$$

The resonance frequency is limited by requirement 5. If this is not valid, it is possible to change the absorbed reactive power and return to Step 2 or change the attenuation and return to Step 4.

6) The quality factor of $L_f - C_f$ series resonant circuit is $10 \leq Q \leq 50$ [18], which can be considered as:

$$Q = \frac{1}{R_f}\sqrt{\frac{L_f}{C_f}} \tag{3.16}$$

where R_f is the gapped equivalent resistance of the inductor L_f.

7) In a real application, the parameter drift of L_f or C_f may result in a resonant frequency variation. Hence, it can be assumed that L_f changes by ±20% while C_f is constant. To ensure that every current harmonic (≥35th) is lower than 0.3% of the fundamental current, it is necessary to analyze the ripple current harmonics both at the switching frequency and at twice the switching frequency while L_f changes by ±20%. If the current harmonics are higher than 0.3%, it is possible to change the absorbed reactive power and return to Step 2 or to change the attenuation and return to Step 4 until the requirement is met.

The design process is iterative for the adjusted *LLCL* parameter values considering the important constraints, such as reactive power, fundamental voltage drop (across the inductors), and amplitude of the harmonic currents.

3.3.2 Saving Analysis on the Grid-Side Inductance

According to the given design criteria, it is reasonable to select the same values of the inverter-side inductors and capacitors in *LCL* and *LLCL* filters. Then, combining (3.13) with (3.14), the value ratio of the grid-side inductors between *LLCL* and *LCL* filters is derived approximately as shown:

$$\gamma(\alpha) = \frac{L_{2\_LLCL}}{L_{2\_LCL}} < \frac{3}{16}\left|\frac{\max\left(|J_1(2\pi\alpha)|,\ |J_3(2\pi\alpha)|,\ |J_5(2\pi\alpha)|\right)}{\max\left(|J_1(\pi\alpha)|,\ |J_3(\pi\alpha)|\right)}\right| \tag{3.17}$$

This is depicted in Figure 3.6 when the modulation index α changes from 0.5 to 1. It can be seen that the value of the grid-side inductor of *LLCL* filter is reduced by a factor of at least 75% compared to that of the *LCL* filter.

3.3.3 Specific Design Consideration for a Simple Passive Damping Strategy

For a small power ratio, especially for the case that uses the current of the inverter-side inductor as the feedback signal, a simple passive damping method is often adopted by inserting a resistor in the harmonic current bypass branch loop to attain improved damping. Then, (3.8) is modified to obtain (3.18):

$$\begin{aligned} G'_{u_i \to i_g}(s) &= \frac{i_g(s)}{u_i(s)}\bigg|_{u_g(s)=0} \\ &= \frac{L_f C_f s^2 + R'_f C_f s + 1}{(L_1 L_2 C_f + (L_1 + L_2) L_f C_f) s^3 + (L_1 + L_2) R'_f C_f s^2 + (L_1 + L_2) s} \end{aligned} \tag{3.18}$$

where R'_f is the sum of the additional damping resistor and R_f.

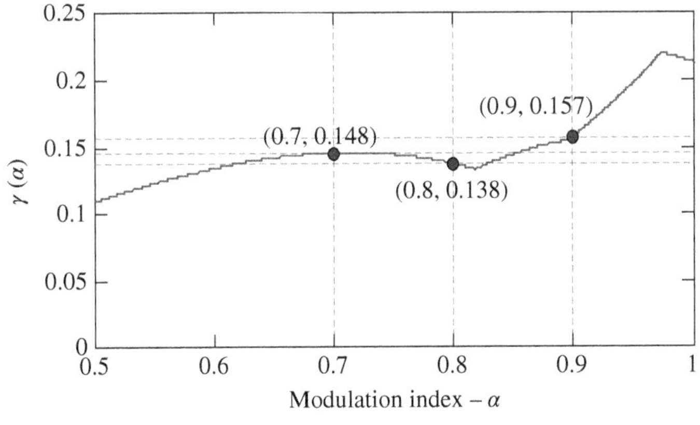

Figure 3.6 Ratio relationship of L_2 between *LLCL* and *LCL* filters.

The design procedure is almost the same except for the value of L_2. When selecting the L_2 of an LCL filter, the only change is to use $G'_{u_i \to i_g}(s)$ instead of $G_{u_i \to i_g}(s)$ in (3.13) with $L_f = 0$. When considering the L_2 of an $LLCL$ filter, it is necessary to not only update (3.14) with $G'_{u_i \to i_g}(s)$ but also to verify whether the harmonic currents around the switching frequency to meet the requirement using the updated (3.13).

3.4 Design Examples for *LCL* and *LLCL* filters

A step-by-step procedure to obtain the values for the parameters of the filter with the following values of $f_s = 20$ kHz, $U_{dc} = 350$ V, $U_g = 220$ V/50 Hz, $P_{rated} = 1.8$ kW, and $\Delta I_1 = 31.54\%$ I_{ref} (the base impedance is 26.89 Ω and upper-limit base capacitance C is 5.92 μF) under unipolar modulation is described as follows:

1) Adopting 1.4% impedance for the inverter-side inductor L_1, a current ripple of 31.54% can be obtained. Finally, the inverter-side inductance is selected as 1.2 mH.
2) The capacitance of C_f is chosen as 2 μF to limit the reactive power to 1.69%. If some constraints cannot be met, C_f should be increased until the limit of 5%. Then, the inductance of L_f in the $L_f - C_f$ series resonant circuit can also be determined.
3) The grid-side inductance of L_2 is easy to calculate. For an LCL filter, L_2 should be equal to the inverter-side inductance of L_1 or smaller for stability reasons [17, 19]. Then, L_2 is selected as 1.2 mH for the LCL filter and 0.22 mH for LLCL filter.
4) The consequent resonance frequency is 4.60 and 7.67 kHz for *LCL* and *LLCL* filters, respectively, which is exactly lower than 10 kHz, as one half of the switching frequency.
5) The quality factor of $L_f - C_f$ series resonant circuit is chosen as 20, and the equivalent resistance of R_f is 0.2 Ω.
6) When the parameter drift of L_f is within ±20%, in order to ensure that each current harmonic (≥35th) is lower than 0.3% of the fundamental current, it is necessary to analyze the ripple current harmonics both at the switching frequency and at twice the switching frequency, as shown in Figure 3.7 and listed in Table 3.1.

The designed parameters of the power filters are listed in Table 3.2, and the grid-side current ripple reduction at the switching frequency and twice the switching frequency is depicted in Figure 3.8. It can be seen that the designed *LLCL* filter has

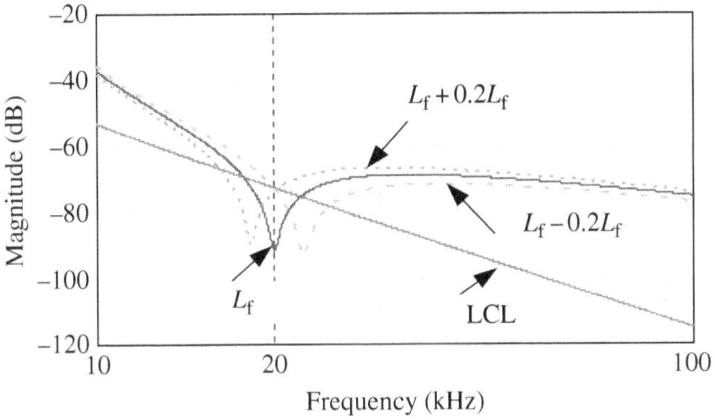

Figure 3.7 Sensitivity analysis of L_f parameter drifting.

Table 3.1 Current harmonics with different L_f.

L_f (μH)	Max harmonics around ω_s (%)	Max harmonics around $2\omega_s$ (%)
26	0.18	0.09
32	0.02	0.14
38	0.18	0.12

Table 3.2 Values of filters in experiments.

Parameters	*LCL* values	*LLCL* values
L_1	1.2 mH	1.2 mH
R_1	0.1 Ω	0.1 Ω
L_2	1.2 mH	0.22 mH
R_2	0.1 Ω	0.05 Ω
L_f	—	32 μH
R_f	—	0.2 Ω
C_f	2 μF	2 μF

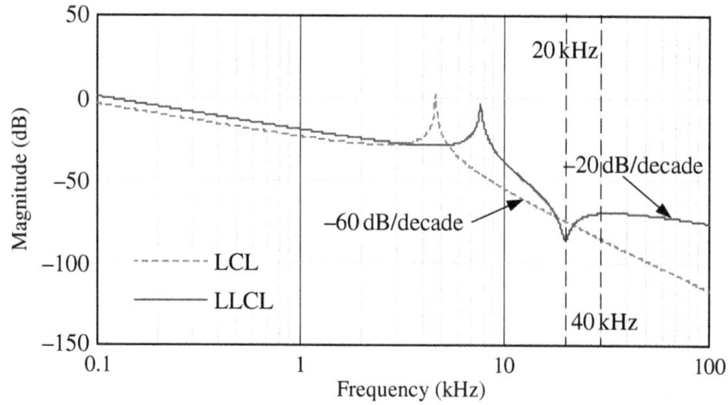

Figure 3.8 Bode plots of transfer functions from u_i to i_g for *LCL* and *LLCL* filters.

better attenuating effects on the harmonics than *LCL* filter around the switching frequency; however, it has opposite effects around twice the switching frequency.

3.5 Experimental Results

To confirm the theoretical analysis further, a 1.8 kW prototype based on DSP (TMS320LF2407A) controller is constructed and the circuit configuration is shown in Figure 3.9. The parameters of the filters are listed in Table 3.2, and the performance evaluation of the inverter with *LLCL* filter is evaluated and investigated under the given conditions of $f_s = 20\,\text{kHz}$, $U_{dc} = 350\,\text{V}$, $U_g = 220\,\text{V}/50\,\text{Hz}$, $P_{rated} = 1.8\,\text{kW}$, grid impedance $Z_g = 0$, and dead time $t_d = 2\,\mu\text{s}$. A programmable ac source (Chroma 6530) is used to emulate the grid voltage.

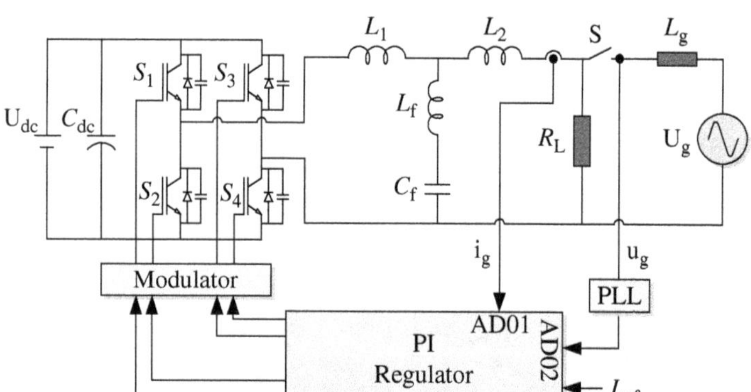

Figure 3.9 Prototype system configuration.

3.5.1 Experimental Results

Referring to the design procedure described in the previous section, experimental analysis is conducted for three cases. Figures 3.10–3.12 show the experimental values of currents and spectra of the three cases, respectively. Case I is the *LCL* filter strategy with $L_2 = 1.2$ mH. Case II is the *LCL* filter strategy with $L_2 = 0.22$ mH. Case III is the *LLCL* filter strategy.

The measured total harmonic distortion (THD)% of i_g in the three cases are listed in Table 3.3, highlighting that the grid-side current in Case III has the lowest THD (1.9%). The grid-side current spectra in the three cases are shown in Figures 3.10b, 3.11b, and 3.12b, respectively. The measured dominant sideband current harmonics, expressed as a percentage of the fundamental current amplitude, are listed in Table 3.3.

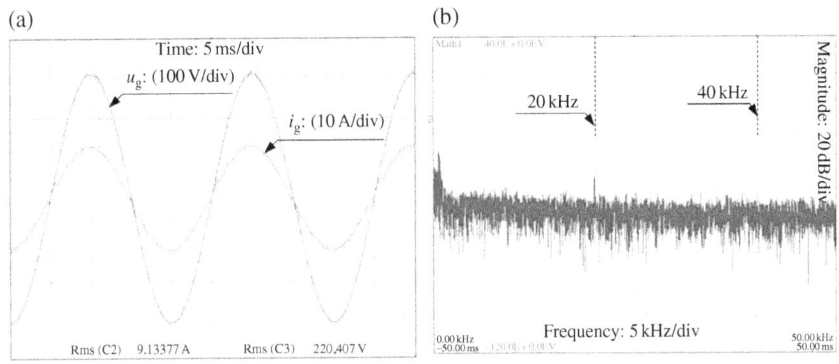

Figure 3.10 Experimental waveforms with *LCL* filter in Case I. (a) Grid voltage and grid-side current waveforms; (b) power spectrum of grid-side current.

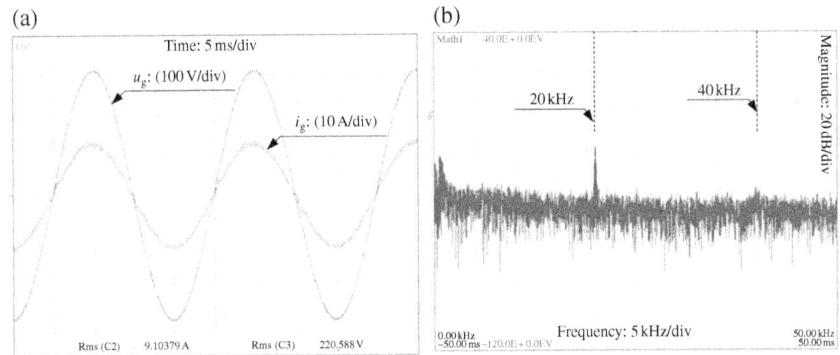

Figure 3.11 Experimental waveforms with *LCL* filter in Case II. (a) Grid voltage and grid-side current waveforms; (b) power spectrum of grid-side current.

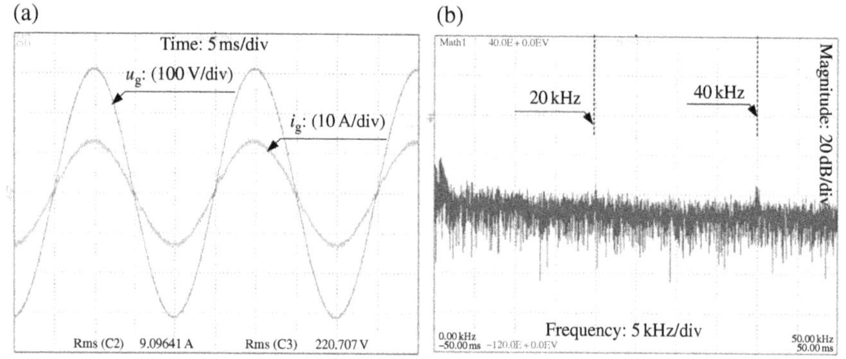

Figure 3.12 Experimental waveforms with *LLCL* filter in Case III. (a) Grid voltage and grid-side current waveforms; (b) power spectrum of grid-side current.

Table 3.3 THD% and dominant sideband harmonic's amplitude of grid-side current in different cases.

	Case I (%)	Case II (%)	Case III (%)
THD (%)	3.4	4.7	1.9
Dominant sideband current harmonic's amplitude	0.28	1.3	0.16

3.5.2 Analysis and Discussion

From the experimental results, the following observations are made.

1) The dominating harmonics current appears at approximately 20 kHz in Cases I and II while in Case III, it appears at approximately 40 kHz. This is consistent with the theoretical analysis.
2) In Case I, the dominating harmonic current in the experiment slightly exceeds the recommendation of IEEE 519-1992 while in Case III, the value meets the recommendation of IEEE 519-1992.
3) The grid-side inductance of *LLCL* filter is reduced by a factor of 81.67% compared to that of *LCL* filter. This is almost the same as the theoretical analysis.
4) In general, the experimental results are consistent with the theoretical analysis especially with regard to the harmonic current attenuation around the switching frequency and twice the switching frequency.

3.6 Summary

In this chapter, the principle of the conventional *LCL* filter was presented. It was observed that the main high-order harmonic currents mostly appear around the switching frequency. Based on this, a novel topology of low-pass power filter with $L_f - C_f$ series resonant circuit known as *LLCL* filter was proposed. In contrast to the *LCL* filter, the *LLCL* filter has nearly zero impedance at the switching frequency and can strongly attenuate the harmonic currents around the switching frequency.

The basic criteria for selecting the parameters and a parametric design procedure of an *LLCL* filter was described. It can be seen that the value of grid-side inductor L_2 of the *LLCL* filter is decided by the harmonic currents around twice the switching frequency instead of those around the switching frequency. This results in a decrease in the total inductance and size reduction compared to the *LCL* filter.

A 1.8 kW single-phase grid-connected inverter was designed to compare the characteristics of the conventional *LCL* filter and proposed *LLCL* filter. Experimental results show that the value of grid-side inductor L_2 of the *LLCL* filter reduced by a factor of 81.67% compared to that of the *LCL* filter when the modulation index was 0.9.

References

1 Blaabjerg, F., Teodorescu, T., Liserre, M., and Timbus, A.V. (2006). Overview of control and grid synchronization for distributed power generation systems. *IEEE Trans. Ind. Electron.* 53 (5): 1398–1409.

2 Salas, V. and Olías, E. (2011). Overview of the state of technique for PV inverters used in low voltage grid-connected PV systems: inverters above 10 kW. *Renew. Sustain. Energy Rev.* 15 (2): 1025–1257.

3 Liserre, M., Teodorescu, R., and Blaabjerg, F. (2006). Stability of photovoltaic and wind turbine grid-connected inverters for a large set of grid impedance values. *IEEE Trans. Power Electron.* 21 (1): 263–272.

4 Wu, W. and Tang, T. (2007). Dual mode time-sharing cascaded sinusoidal inverter. *IEEE Trans. Energy Convers.* 22 (3): 795–797.

5 Ahmed, N.A., Woo Lee, H., and Nakaoka, M. (2007). Dual-mode time-sharing sinewave-modulation soft switching boost full-bridge one-stage power conditioner without electrolytic capacitor DC link. *IEEE Trans. Ind. Appl.* 43 (3): 805–813.

6 Shen, J., Rigbers, K., Dick, C.P., and De Doncker, R.W. (2008). A dynamic boost converter input stage for a double 120 flattop modulation based three-phase inverter.

Proceedings of the IEEE Industry Applications Society, pp. 1–7, Edmonton, AB, Canada (5–8 October 2008). IEEE.

7 Lindgren, M. and Svensson, J. (1995). Connecting fast switching voltage source converters to the grid – harmonic distortion and its reduction. *IEEE Strock PowerTech Conference*, pp. 191–195, Stockholm, Sweden (18–22 June 1995).

8 Gabe, I.J., Montagner, V.F., and Pinheiro, H. (2009). Design and implementation of a robust current controller for VSI connected to the grid through an LCL filter. *IEEE Trans. Power Electron.* 24 (6): 1444–1452.

9 Jeong, H.-G., Lee, K.-B., Choi, S., and Choi, W. (2010). Performance improvement of LCL-filter-based grid-connected inverters using PQR power transformation. *IEEE Trans. Power Electron.* 25 (5): 1320–1330.

10 Dannehl, J., Fuchs, F.W., and Thøgersen, P.B. (2010). PI state space current control of grid-connected PWM converters with LCL filters. *IEEE Trans. Power Electron.* 25 (9): 2320–2330.

11 Channegowda, P. and John, V. (2010). Filter optimization for grid interactive voltage source inverters. *IEEE Trans. Ind. Electron.* 57 (12): 4106–4114.

12 Liserre, M., Blaabjerg, F., and Hansen, S. (2005). Design and control of an LCL-filter-based three-phase active rectifier. *IEEE Trans. Ind. Appl.* 41 (5): 1281–1291.

13 Holmes, D.G. and Lipo, T.A. (2003). *Pulse Width Modulation for Power Converters: Principles and Practice*, 183–199. Hoboken, NJ: Wiley.

14 IEEE Application Guide for IEEE Std. 1547 (2008). *IEEE Standard for Interconnecting Distributed Resources with Electric Power Systems.* IEEE Standard 1547.2-2008.

15 IEEE (1992). Recommended practices and requirements for harmonic control in electrical power systems. *519-2014 – IEEE Recommended Practice and Requirements for Harmonic Control in Electric Power Systems.*

16 M. Liserre, F. Blaabjerg, and A. Dell'Aquila, "Step-by-step design procedure for a grid-connected three-phase PWM voltage source converter," Int. J. Electron., vol. 91, no. 8, pp. 445–460, Aug. 2004.

17 Lang, Y., Xu, D., Hadianamrei, S.R., and Ma, H. (2005). A novel design method of LCL type utility interface for three-phase voltage source rectifier. *Proceedings of the IEEE Power Electronics Specialists Conference*, pp. 313–317, Recife, Brazil (16 June 2005). New York: IEEE.

18 Phipps, J.K. (1997). A transfer function approach to harmonic filter design. *IEEE Ind. Appl. Mag.* 3 (2): 68–82.

19 Bolsens, B., De Brabendere, K., Vnan den Keybus, J. et al. (2006). Model-based generation of low distortion currents in grid-coupled PWM-inverters using an LCL output filter. *IEEE Trans. Power Electron.* 21 (4): 1032–1040.

4

Modeling and Suppressing Conducted Electromagnetic Interference Noise for LCL/LLCL-Filtered Single-Phase Transformerless Grid-Connected Inverter

4.1 Introduction

The characteristic scheme of the traditional *LCL* filter and proposed *LLCL* filter were analyzed in the previous chapter. It was observed that the *LLCL* filter can improve the suppression of switching harmonics by connecting a resonant inductor in series with the filter capacitor circuit of the traditional *LCL* filter to reduce the total inductance. Nevertheless, the ability of the system to suppress conducted electromagnetic interference (EMI) noise is reduced owing to the additional resonant inductor. Additionally, its common mode (CM) and differential mode (DM) EMI noise attenuation cannot meet the requirements of *Comité International Spécial des Perturbations Radioélectriques* (CISPR) Class A. Therefore, the focus of this chapter is the effective suppression of voltage noise through improvements in topology.

The conducted EMI noise can be divided into two categories: CM and DM. For single-phase grid-connected inverters (GCI) using unipolar power width modulation, both CM and DM EMI noise must be effectively suppressed to meet the electromagnetic compatibility standards. Similarly, for single-phase GCI with bipolar modulation or special GCI with CM EMI noise suppression capabilities, it is necessary to effectively suppress the DM EMI noise to meet the corresponding electromagnetic compatibility standards.

In this chapter, the characteristics of conducted EMI noise for single-phase transformerless GCI with *LLCL*/*LCL* filter is described in detail in Section 4.2.

In Section 4.3, an effective measure for EMI suppression is proposed for the single-phase transformerless GCI with an *LLCL* filter using discontinuous unipolar pulse width modulation (DUPWM). Compared to the traditional *LLCL* filter, this method involves inserting two additional shunt capacitors in parallel in the resonant circuit, and the midpoint of the branch of the additional shunt capacitor is connected to that on the DC side. Thus, most of the CM EMI noise can be bypassed by

Control and Filter Design of Single-Phase Grid-Connected Converters, First Edition. Weimin Wu, Frede Blaabjerg, Henry Chung, Yuanbin He, and Min Huang.

Published 2023 by John Wiley & Sons, Inc.

additional parallel capacitors, and the ability of the system to suppress DM EMI noise increases thereby suppressing the conducted EMI noise and reducing the leakage current when using DUPWM.

Since the single-phase full-bridge GCI uses bipolar pulse width modulation (BPWM), the CM EMI noise can be ignored. Therefore, the system only needs to suppress the DM EMI noise. We have proposed a DM EMI noise suppression method in Section 4.4 to minimize the additional reactive power. In this method, an LC resonant circuit at 150 kHz is connected in parallel at both ends of the notch filter branch circuit of the traditional *LLCL* filter.

Additionally, a 500 W single-phase full-bridge GCI prototype with a DSP controller has been developed. The evaluation and research experiments were conducted under the given conditions of $f_s = 20$ kHz, $U_{dc} = 175$ V, $U_g = 100$ V/50 Hz, and $P_{rated} = 500$ W. The theoretical analysis and experimental results indicate that, unlike the traditional *LLCL* filter, both the improved *LLCL* filter solutions can meet the standards of CISPR Class A, and thereby effectively suppress the conducted EMI noise of the single-phase transformerless GCI.

4.2 Conducted EMI Noise Analysis

4.2.1 CM and DM Voltage Noises

Figure 4.1 shows the configuration of a single-phase full-bridge GCI with the conventional *LLCL*-filter where the stray capacitor C_p in the PV applications is considered, and a simplified line impedance stabilization network (LISN) module

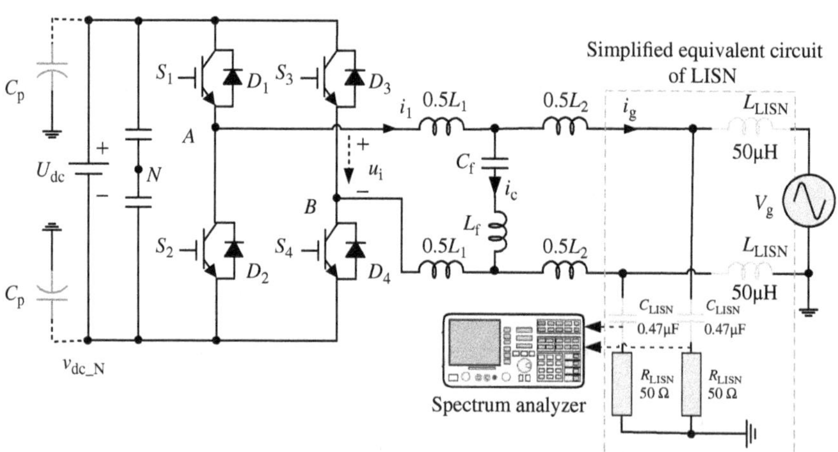

Figure 4.1 Conventional *LLCL*-Filtered single-phase GCI connected to a line impedance stabilization network (LISN).

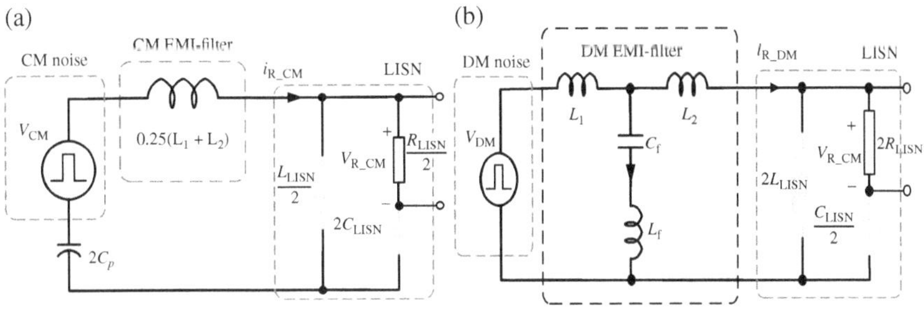

Figure 4.2 Equivalent circuit of conventional *LLCL*-filtered system for analyzing (a) common mode (CM) voltage noise; (b) differential mode (DM) voltage noise.

is used as an interface between the inverter and grid to measure the EMI noise. For analyzing CM and DM EMI noises, the ideal equivalent circuits of the conventional *LLCL*-filtered inverter system are illustrated in Figure 4.2.

The components of CM voltage $V_{CM}(t)$ and DM voltage $V_{DM}(t)$ can be calculated as:

$$V_{CM}(t) = \frac{V_{AN}(t) + V_{BN}(t)}{2}, V_{DM}(t) = V_{AN}(t) - V_{BN}(t) \tag{4.1}$$

where $V_{AN}(t)$ and $V_{BN}(t)$ are the terminal voltages of the two phase legs with respect to the midpoint of the split dc capacitors N as shown in Figure 4.1.

Within the frequency range of 150 kHz–1 MHz, the attenuation gains of CM and DM voltage noises for the ideal conventional *LLCL* filter can be approximately derived using (4.2) and (4.3), respectively, as:

$$\begin{aligned} Att'_{CM\_LLCL}(\omega)|\mathrm{dB}|\Big|_{\omega \geq 2\pi \cdot 150\ \mathrm{kHz}} &\approx -20\log_{10}\Big|0.25j\omega(L_1 + L_2) \\ &+ \frac{1}{2j\omega C_P} + Z_{LISN\_1}\Big| + 20\log_{10}\left|\frac{j\omega L_{LISN}}{2}\right| \\ &- 20\log_{10}\left|\frac{R_{LISN}}{2} + \frac{1}{2j\omega C_{LISN}} + \frac{j\omega L_{LISN}}{2}\right| \\ &+ 20\log_{10}\frac{R_{LISN}}{2} \end{aligned} \tag{4.2}$$

$$\begin{aligned} Att'_{DM\_LLCL}(\omega)|\mathrm{dB}|\Big|_{\omega \geq 2\pi \cdot 150\ \mathrm{kHz}} &\approx -20\log_{10}\left|(j\omega L_2 + Z_{LISN\_2})//Z_f + j\omega L_1\right| \\ &+ 20\log_{10}(2R_{LISN}) + 20\log_{10}\left|\frac{Z_f}{j\omega L_2 + Z_f + Z_{LISN\_2}}\right| \\ &+ 20\log_{10}\left|\frac{j\omega L_{LISN}}{R_{LISN} + 1/j\omega C_{LISN} + j\omega L_{LISN}}\right| \end{aligned} \tag{4.3}$$

where $Z_{LISN\_1} = 0.5j\omega L_{LISN}//(0.5R_{LISN} + 1/2\,j\omega C_{LISN})$, $Z_{LISN\_2} = 2j\omega L_{LISN}//(2R_{LISN} + 2/j\omega C_{LISN})$, and $Z_f = j\omega L_f + 1/j\omega C_f$

As shown in Figure 4.1, if the resonant inductor L_f is shortened, a conventional *LCL*-filtered system is obtained. The ideal equivalent circuit of the conventional *LCL*-filtered system is similar to that of conventional *LLCL* filter to analyze the CM voltage noise, which is shown in Figure 4.2a. The related attenuation gain of the CM voltage noise is highlighted in (4.2). As shown in Figure 4.2b, if the resonant inductor L_f is set to zero, the equivalent circuit of the conventional *LCL*-filtered system for analyzing DM voltage noise can also be obtained. Then, the attenuation gain of the conducted DM voltage noise through the ideal conventional *LCL* filter can be calculated as:

$$\begin{aligned} Att'_{DM\_LCL}(\omega)|\mathrm{dB}|\Big|_{\omega \geq 2\pi\cdot 150\ \mathrm{kHz}} &\approx -20\log_{10}\left|(j\omega L_2 + Z_{LISN\_2})//1/j\omega C_f + j\omega L_1\right| \\ &+ 20\log_{10}(2R_{LISN}) + 20\log_{10}\left|\frac{1/j\omega C_f}{j\omega L_2 + 1/j\omega C_f + Z_{LISN\_2}}\right| \\ &+ 20\log_{10}\left|\frac{j\omega L_{LISN}}{R_{LISN} + 1/j\omega C_{LISN} + j\omega L_{LISN}}\right| \end{aligned} \tag{4.4}$$

4.2.2 Spectrum of DM and CM Voltage Noise for GCI Using DUPWM

For single-phase full-bridge inverter applications, the DUPWM method is extensively used because it causes less switching power losses. When the asymmetrical and regularly sampled DUPWM method [1] is adopted, the spectrum of DM and CM voltage noise generated by an ideal full-bridge single-phase inverter can be expressed as:

$$v_{DM}(t) = \alpha U_{dc}\cos(\omega_o t) + \frac{2U_{dc}}{\pi}\sum_{m=1}^{\infty}\sum_{n=-\infty}^{\infty}\frac{1}{m}J_n(m\pi\alpha)\sin\frac{n\pi}{2}\cos(m\omega_n t + n\omega_o t) \tag{4.5}$$

$$\begin{aligned} v_{CM}(t) = {} & \frac{\alpha U_{dc}}{\pi} - 0.5\alpha U_{dc}\cos(\omega_o t) - \frac{2\alpha U_{dc}}{\pi}\sum_{n=2}^{\infty}\frac{\cos n\pi/2}{(n+1)(n-1)}\cos(n\omega_o t) \\ & + \frac{4U_{dc}}{\pi^2}\sum_{m=1}^{\infty}\frac{1}{m}\sum_{k=1}^{\infty}\frac{1}{2k-1}J_{2k\text{-}1}(m\pi\alpha)\cos(m\omega_s t) \\ & - \frac{U_{dc}}{\pi}\sum_{m=1}^{\infty}\frac{1}{m}\sum_{k=-\infty}^{\infty}J_n(m\pi\alpha)\sin\frac{n\pi}{2}\cos(m\omega_s t + n\omega_o t) \\ & - \frac{4U_{dc}}{\pi^2}\sum_{m=1}^{\infty}\frac{1}{m}\sum_{\substack{n=-\infty \\ (n\neq 0)}}^{\infty}\sum_{k=1}^{\infty}J_{2k\text{-}1}(m\pi\alpha)\frac{(2k-1)\cos n\pi/2}{(n+2k-1)(n-2k+1)}\Big|_{|n|\neq 2k-1}\cos(m\omega_s t + n\omega_o t) \end{aligned} \tag{4.6}$$

Table 4.1 Parameters of *LCL* and *LLCL*-filtered inverters using DUPWM at a power level of 500 W/110 V.

Parameters	Conventional *LCL* filter	Conventional *LLCL* filter	Modified *LCL* filter	Modified *LLCL* filter
L_1	1.2 mH	1.2 mH	1.2 mH	1.2 mH
L_2	1.2 mH	0.22 mH	0.6 mH	0.18 mH
C_f	2 μF	2 μF	2 μF	2 μF
L_f	—	32 μH	—	32 μH
C_{CM}	—	—	4 μF	4 μF
C_P	44 nF	44 nF	44 nF	44 nF

where α is the modulation index, U_{dc} is the DC-link voltage, and J_n is the Bessel function [1]. ω_o and ω_s are the fundamental and switching frequencies in radians per second, respectively.

During the EMI noise analysis, the total EMI noise should not only be related to the background EMI noise caused by the auxiliary power supply of the controller but also on the parasitic parameters, which are closely related to the circuit layout and character of the switches [2–4]. However, according to [2, 5], the EMI effect caused by the parasitic parameters of the output filter in the frequency range of 150 kHz–1 MHz is not significant. Therefore, to further illustrate the EMI noise for a high-order power-filtered system within the same frequency range, simulations are performed. The parameters for simulation are listed in Table 4.1 under the following conditions: U_{dc} = 175 V, U_g = 110 V/50 Hz, P_{rated} = 500 W, and switching frequency is 20 kHz. The parameters of the filters depend on the harmonic current requirements of IEEE1547.2-2008 and 519-1992. The corresponding design procedure has been discussed in [6].

Under ideal conditions and within the frequency range of 150 kHz–1 MHz, the simulated maximum amplitude of CM and DM voltage noises for conventional *LLCL* filter and *LCL* filter-based single-phase full-bridge GCI using DUPWM method are shown in Figure 4.3. It can be seen that for the conventional *LLCL* filter-based system, both the simulated attenuations for CM and DM voltage noises cannot meet the standards of CISPR Class A or Class B whereas for conventional *LCL* filter, CM voltage noise cannot meet the EMI standards. Therefore, for a conventional high-order power filter-based system, further measures should be taken to ensure that the conducted EMI noise meets the requirements of CISPR standards.

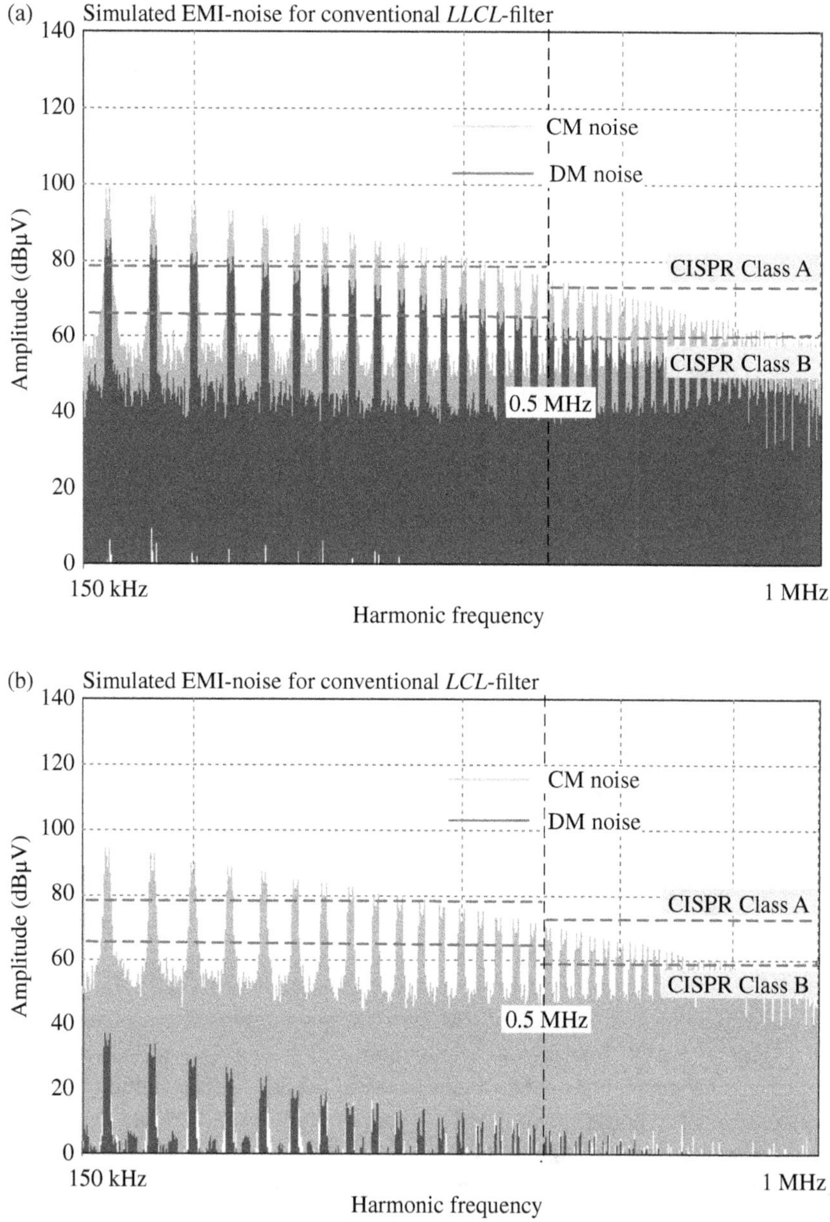

Figure 4.3 Simulated ac-side conducted EMI voltage noise of (a) conventional *LLCL*-filtered GCI; (b) conventional *LCL*-filtered GCI.

4.2.3 Spectrum of DM Voltage Noise for GCI Using BPWM

When the asymmetrical and regularly sampled bipolar SPWM method [1] is adopted, the spectrum of DM voltage generated by an ideal full-bridge single-phase inverter can be expressed using (4.7) as:

$$
\begin{aligned}
v_{\mathrm{DM}}(t) = {} & \alpha U_{\mathrm{dc}} \sin(\omega_0 t) \\
& + \frac{4U_{\mathrm{dc}}}{\pi} \sum_{n=1}^{\infty} \frac{1}{n} J_0\left(\frac{\alpha n\pi}{2}\right) \sin\frac{n\pi}{2} \cos(n\omega_0 t) \\
& + \frac{8U_{\mathrm{dc}}}{\pi} \sum_{n=1}^{\infty} \frac{1}{n} \sum_{m=1}^{\infty} \frac{1}{m} J_{2m\text{-}1}\left(\frac{\alpha n\pi}{2}\right) \sin|(2m-1)\omega_0 t| \cos\frac{n\pi}{2} \cos(n\omega_0 t) \\
& + \frac{8U_{\mathrm{dc}}}{\pi} \sum_{n=1}^{\infty} \frac{1}{n} \sum_{m=1}^{\infty} \frac{1}{m} J_{2m\text{-}1}\left(\frac{\alpha n\pi}{2}\right) \cos|2m\omega_0 t| \sin\frac{n\pi}{2} \cos(n\omega_0 t)
\end{aligned} \tag{4.7}
$$

According to a similar power filter design method introduced in [6], the parameters of *LCL* and *LLCL* filters in BPWM method can be calculated as listed in Table 4.2 under the conditions of $U_{\mathrm{dc}} = 175$ V, $U_{\mathrm{g}} = 110$ V/50 Hz, $P_{\mathrm{rated}} = 500$ W, and switching frequency of 20 kHz.

Theoretically, the EMI noise depends significantly on the parasitic parameters of the inverter system; however, according to [2, 5, 7], the parasitic parameters have a negligible effect on the measured conducted DM EMI noise within the frequency range of 150 kHz–1 MHz. To further illustrate the DM EMI noise characteristics, simulations are performed for the *LCL* and *LLCL* filter-based GCIs within the frequency range of 150 kHz–1 MHz.

Figure 4.4 shows the simulated DM EMI noise for the conventional *LCL* filter and *LLCL*-filtered GCIs. It can be seen that, for conventional *LLCL*-filtered GCI within the frequency range of 150–300 kHz, the EMI requirements of CISPR11 Class A cannot be met while for the *LCL*-filtered GCI, there is at least a margin of 40 dB/μV to meet the requirements of CISPR 11 Class A.

Table 4.2 Parameters of *LCL* and *LLCL*-filtered inverters using BPWM at a power level of 500 W/110 V.

Parameters	L_1	C_f	L_2	L_f
LCL filter	2.4 mH	2 μF	1.6 mH	0
LLCL filter	2.4 mH	2 μF	0.22 mH	32 μH

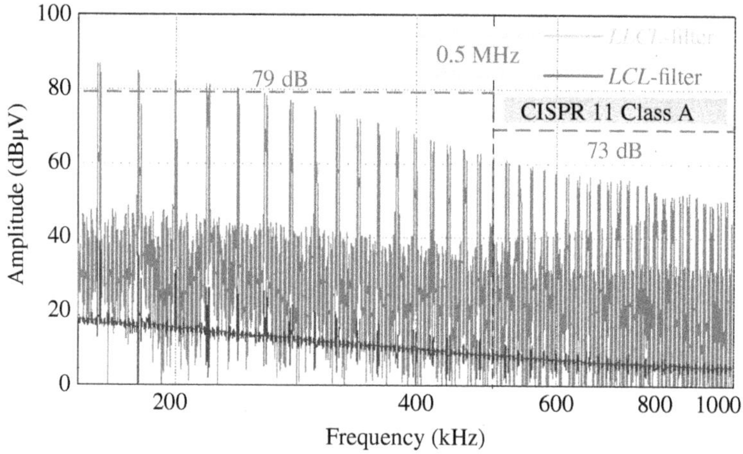

Figure 4.4 Simulated DM EMI noise for conventional *LCL* and *LLCL*-filtered GCIs including CISPR11 Class A limits.

4.3 Modified LLCL Filter to Fully Suppress the Conducted EMI Noise for GCI Using DUPWM

In Section 4.2, the EMI voltage noise characteristics of *LLCL*-filtered GCI was described in detail. For a single-phase full-bridge transformerless GCI, the CM loop current is smaller than the DM loop current. When using BPWM, the impact of CM EMI noise can be ignored. However, for inverters using DUPWM, CM EMI noise cannot be ignored resulting in failure to meet the CISPR Class A standards. Therefore, to ensure that the EMI noise of the *LLCL*-filtered system meets the requirements, we have proposed a novel solution in this section that uses parallel split capacitors to reduce the EMI noise.

4.3.1 Modified Solution for *LLCL* Filter

Dong et al. [8] proposed an interesting method to suppress the leakage current of an *LCL*-filtered single-phase GCI system. In this chapter, a similar structure is proposed for the *LLCL*-filtered GCI using DUPWM as shown in Figure 4.5. Compared to the conventional *LLCL* filter, two extra split CM capacitors C_{CM} are connected in parallel with the L_f–C_f resonance circuit. The midpoint of the extra split CM capacitors is linked to the midpoint of the split DC capacitors C_{dc_s}. Then, most of the high-frequency CM EMI noise currents pass through the split capacitor branch. Simultaneously, the series noninductive split CM capacitors also suppress the DM EMI noise.

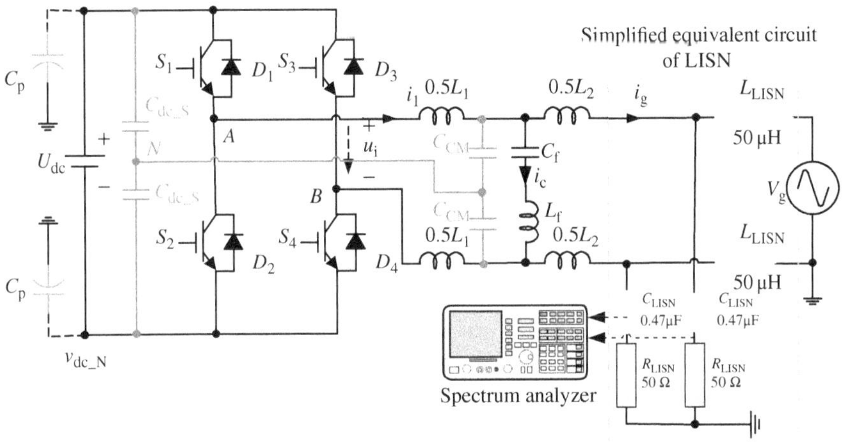

Figure 4.5 Proposed modified *LLCL*-filtered GCI using DUPWM and measuring system.

For the modified *LLCL*-filtered system, the ideal equivalent circuits for analyzing the CM and DM voltage noises are illustrated in Figure 4.6. The attenuation gains for suppressing the CM and DM voltage noise can be approximately calculated as:

$$\begin{aligned}
Att'_{\mathrm{CM_LLCL}}(\omega)|\mathrm{dB}|\Big|_{\omega \geq 2\pi\cdot 150\ \mathrm{kHz}} \approx &-20\log_{10}\left|\left(0.25j\omega L_2 + Z_{\mathrm{LISN_1}} + 1/2j\omega C_p\right)\right. \\
&\left.//1/j\omega C_s + 0.25j\omega L_1\right| \\
&-20\log_{10}|0.25j\omega L_2 + Z_{\mathrm{LISN_1}} + 1/2j\omega C_p + 1/j\omega C_s| \\
&-20\log_{10}|j\omega C_s| \\
&-20\log_{10}|0.5j\omega L_{\mathrm{LISN}} + 1/2j\omega C_{\mathrm{LISN}} + 0.5R_{\mathrm{LISN}}| \\
&+20\log_{10}\frac{R_{\mathrm{LISN}}}{2} + 20\log_{10}\left|\frac{j\omega L_{\mathrm{LISN}}}{2}\right|
\end{aligned} \tag{4.8}$$

$$\begin{aligned}
Att'_{\mathrm{DM_LLCL}}(\omega)|\mathrm{dB}|\Big|_{\omega \geq 2\pi\cdot 150\ \mathrm{kHz}} \approx &-20\log_{10}\left|(j\omega L_2 + Z_{\mathrm{LISN_1}})//Z_{\mathrm{f_1}} + j\omega L_1\right| \\
&+20\log_{10}(2R_{\mathrm{LISN}}) \\
&+20\log_{10}\left|\frac{Z_{\mathrm{f_1}}}{j\omega L_2 + Z_{\mathrm{f_1}} + Z_{\mathrm{LISN_2}}}\right| \\
&+20\log_{10}\left|\frac{j\omega L_{\mathrm{LISN}}}{R_{\mathrm{LISN}} + 1/j\omega C_{\mathrm{LISN}} + j\omega L_{\mathrm{LISN}}}\right|
\end{aligned} \tag{4.9}$$

where $C_s = \dfrac{2C_{\mathrm{CM}}C_{\mathrm{dc\_s}}}{C_{\mathrm{CM}} + C_{\mathrm{dc\_s}}}$, $Z_{\mathrm{f\_1}} = \left(j\omega L_f + 1/j\omega C_f\right)//2/j\omega C_{\mathrm{CM}}$.

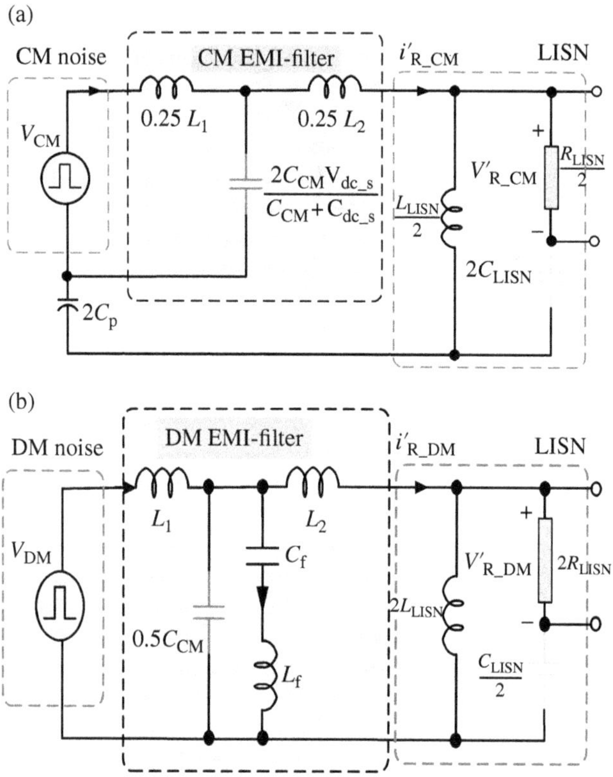

Figure 4.6 Equivalent circuit of the modified *LLCL* filter for analyzing (a) CM voltage noise; (b) DM voltage noise.

As shown in Figure 4.5, modified *LCL*-filtered GCI is obtained when L_f is shortened, which is the same as the filter structure proposed by Dong et al. [8]. The ideal equivalent circuit of the modified *LCL*-filtered system for analyzing the CM voltage noise is also shown in Figure 4.6a, and the attenuation gain of the CM voltage noise can be calculated using (4.8). The equivalent circuit for analyzing the DM voltage noise can be obtained when L_f is shortened as shown in Figure 4.6b, and the related attenuation gain can be depicted as:

$$
\begin{aligned}
Att'_{DM\_LCL}(\omega)|dB|\Big|_{\omega\geq 2\pi\cdot 150\,\text{kHz}} \approx & -20\log_{10}\left|(j\omega L_2+Z_{LISN\_2})//{}^{1}\!/\!{}_{j\omega(C_f+0.5C_{CM})}+j\omega L_1\right| \\
& +20\log_{10}(2R_{LISN}) \\
& +20\log_{10}\left|\frac{{}^{1}\!/\!{}_{j\omega(C_f+0.5C_{CM})}}{j\omega L_2+{}^{1}\!/\!{}_{j\omega(C_f+0.5C_{CM})}+Z_{LISN\_2}}\right| \\
& +20\log_{10}\left|\frac{j\omega L_{LISN}}{R_{LISN}+{}^{1}\!/\!{}_{j\omega C_{LISN}}+j\omega L_{LISN}}\right|
\end{aligned}
\tag{4.10}
$$

Based on the parameters in Table 4.1, the simulated maximum amplitudes of the attenuations of CM and DM voltage noises for the modified *LLCL* and *LCL* filters are shown in Figure 4.7. Compared to Figure 4.3, the attenuation of EMI noise within the frequency range of 150 kHz–1 MHz has significantly improved.

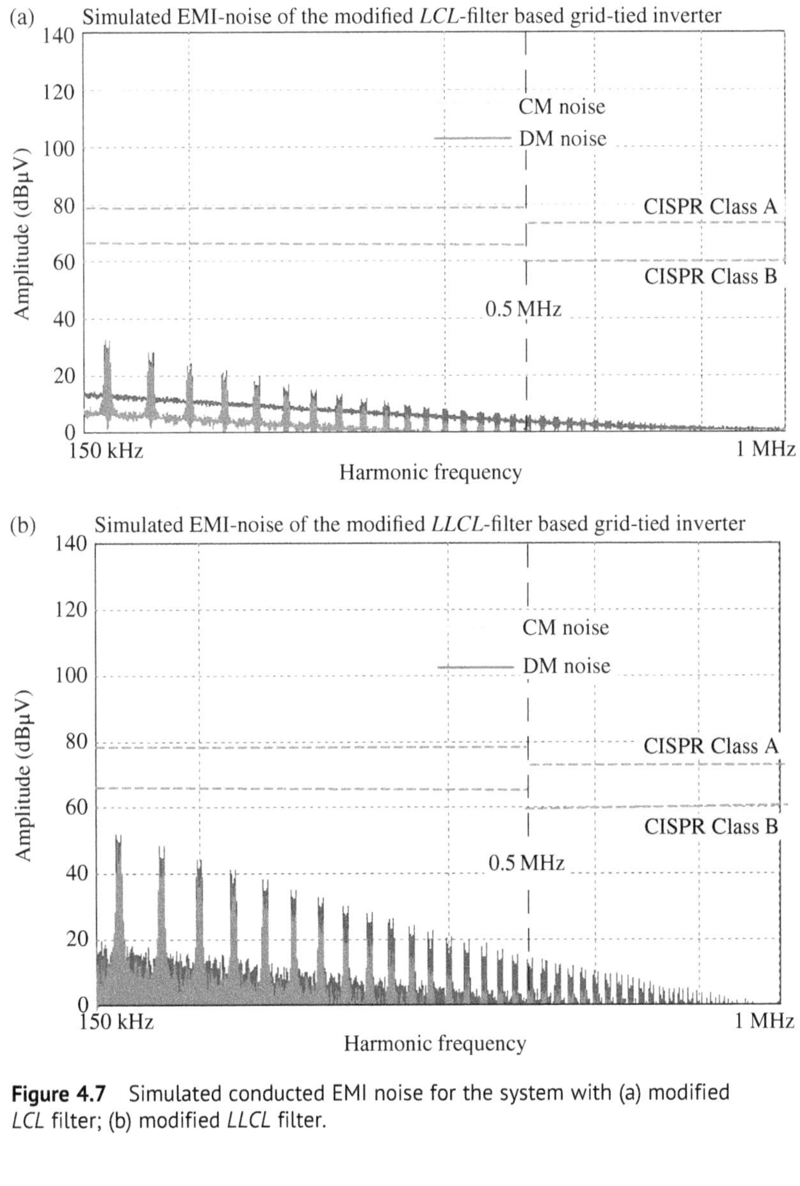

Figure 4.7 Simulated conducted EMI noise for the system with (a) modified *LCL* filter; (b) modified *LLCL* filter.

4.3.2 Improved Parameter Design of *LLCL* filter

To ensure that the improved filter meets the requirements for EMI noise within 150 kHz–1 MHz, we have introduced relevant improvements in the *LLCL* filter. In this section, we have focused on designing corresponding component parameters of improved *LLCL* filter around the additional split capacitor C_M and grid inductor L_2.

4.3.3 Constraints on Harmonics of the Grid-Injected Current and EMI Noise Within 150 kHz to 1 MHz

In [6], a step-by-step design procedure of the conventional *LLCL* filter was discussed based on the requirements of five limits that were also discussed in [9, 10]. For the modified *LLCL* filter, the parameters of the inverter-side inductor and $L_f - C_f$ resonant circuit are similar to those of the conventional *LLCL* filter. In this subsection, we have focused on designing the grid-side inductor and additional split CM capacitor. Certainly, the trial-and-error method was used for the design. The additional split CM capacitor depends on the rule of $0.5C_{CM} + C_f \leq \frac{5\%P_{rated}}{\omega_0 V_g V_q}$ where P_{rated} is the rated output power of the inverter and V_g is the RMS value of the grid voltage. In this subsection, the additional split CM capacitor is selected according to the following equation:

$$C_f = \frac{1.5\%P_{rated}}{\omega_0 V_g V_q} \tag{4.11}$$

The transfer function of the grid-injected current versus output voltage for the modified *LLCL*-filtered inverter can be derived as:

$$\begin{aligned} G_{u_i \to i_g}(s) &= \frac{i_g(s)}{u_i(s)}\bigg|_{s=j\omega} \\ &= \left|\frac{Z_C(s)}{Z_1(s)Z_2(s) + Z_1(s)Z_C(s) + Z_C(s)Z_2(s)}\right|_{s=j\omega} \end{aligned} \tag{4.12}$$

where $Z_1(s) = sL_1$, $Z_2(s) = sL_2$, and $Z_c(s) = \left(1/0.5sC_{CM}\right)//\left(sL_f + 1/sC_f\right)$. Figure 4.8 shows the Bode diagrams of the grid-injected current versus ac output voltage for the *LLCL*-filtered GCI system. It can be seen that the current harmonics at approximately the switching frequency are attenuated to over 80 dB for both conventional *LLCL* filter and modified *LLCL* filter-based system owing to the $L_f - C_f$ resonant circuit. If the harmonics at approximately twice the switching frequency are sufficiently small, then the harmonic current requirements of IEEE519-1992 can be met. Therefore, the criteria to choose grid side L_2 of the modified *LLCL* filter are expressed as:

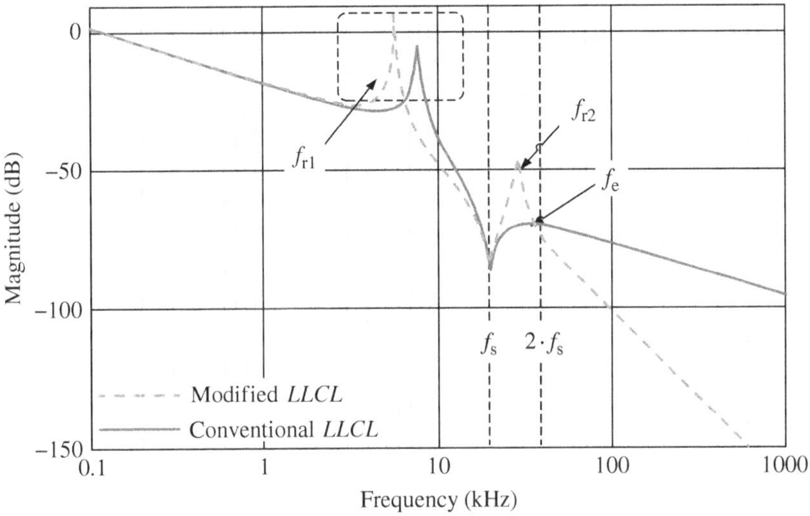

Figure 4.8 Bode diagrams of the grid-injected current versus ac output voltage for *LLCL*-filtered GCI.

$$\frac{{}^{U_{dc}}\!/_{\pi} \left| \max\left(|J_1(2\pi\alpha)|,\ |J_3(2\pi\alpha)|,\ |J_5(2\pi\alpha)| \right) \times \left| G_{u_i \to i_g}(j2\omega_s) \right| \right|}{I_{ref}} \le 0.3\% \tag{4.13}$$

where I_{ref} is the rated reference peak current.

4.3.3.1 Constraints on Leakage Current

In many applications such as PV generation, the inverter generates a variable CM voltage if it is performed without galvanic isolation. Additionally, the leakage current (CM current) appears in the stray capacitor between the PV array and protection earth [11]. Therefore, for safety considerations, the leakage current should be limited to the required level [12].

For the modified *LLCL* filter-based GCI, the equivalent circuit for analyzing the leakage current is shown in Figure 4.9. The split CM capacitors provide a high-frequency attenuation loop on the CM voltage noise as well as the DC leakage current. The negative dc-rail voltage with respect to the earth v_{dc_N} and leakage current can be estimated as:

$$v_{dc\_N} \approx -\frac{v_{CM}(\omega)}{1 - \left({}^{\omega}\!/_{\omega_r}\right)^2} - 0.5V_{dc} + 0.5V_g \tag{4.14}$$

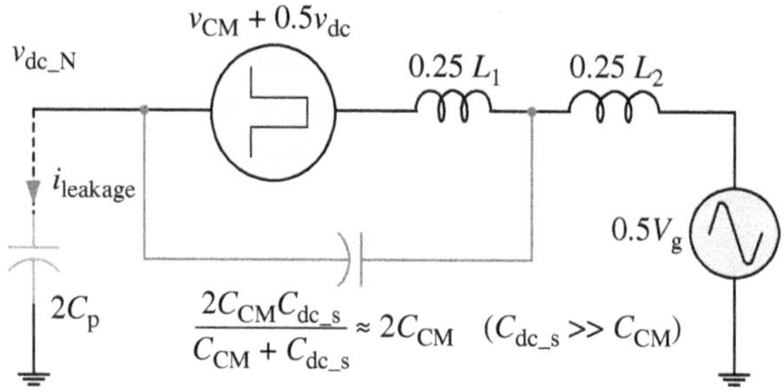

Figure 4.9 Equivalent circuit for analyzing the leakage current of modified *LCL*/*LLCL*-filtered GCI.

$$i_{leakage} \approx \sqrt{\sum_{m=1}^{\infty} \sum_{n=-\infty}^{\infty} \left|2(m\omega_n + n\omega_o) \cdot C_p \cdot v_{dc\_N}(m\omega_n + n\omega_o)\right|^2} \tag{4.15}$$

where $\omega_r = \sqrt{\frac{2(C_{CM} + C_{dc\_N})}{L_1 \cdot C_{CM} \cdot C_{dc\_N}}}$ V_g is the grid voltage, V_{dc} is the DC-link voltage of the inverter, and $i_{leakage}$ is the RMS value of the leakage current.

When CM and L_2 are selected, the EMI requirement within the frequency range of 150 kHz–1 MHz must be verified with (4.8) and (4.9), and the total EMI noise requirement should be met. Additionally, a coupled CM inductor L_{CM} must be connected to the inverter-side inductor to further reduce the CM EMI noise and leakage current.

4.3.4 Experimental Verification

To confirm the effectiveness of the proposed modified *LLCL* filter in suppressing the conducted EMI noise, a 500 W prototype of the single-phase full-bridge GCI using a DSP (TMS320LF2812A) controller was constructed. The experiments were evaluated and investigated under the given conditions of $f_s = 20$ kHz, $U_{dc} = 175$ V, $U_g = 100$ V/50 Hz, and $P_{rated} = 500$ W. DUPWM method was adopted for the analysis. The experimental parameters of the filter are the same as those used for the simulations as listed in Table 4.1.

A photo of the modified *LLCL*-filtered inverter system is shown in Figure 4.10. Note that the size of L_2 is significantly smaller than L_1. Certainly, further optimization of the design of the total filter is essential to minimize the total size and footprint.

The experimental results of the power spectrum of grid-injected current, conducted EMI noise on the grid-side, negative dc-rail voltage with respect to the earth, V_{dc_N}, and leakage current of the four filter cases are described below.

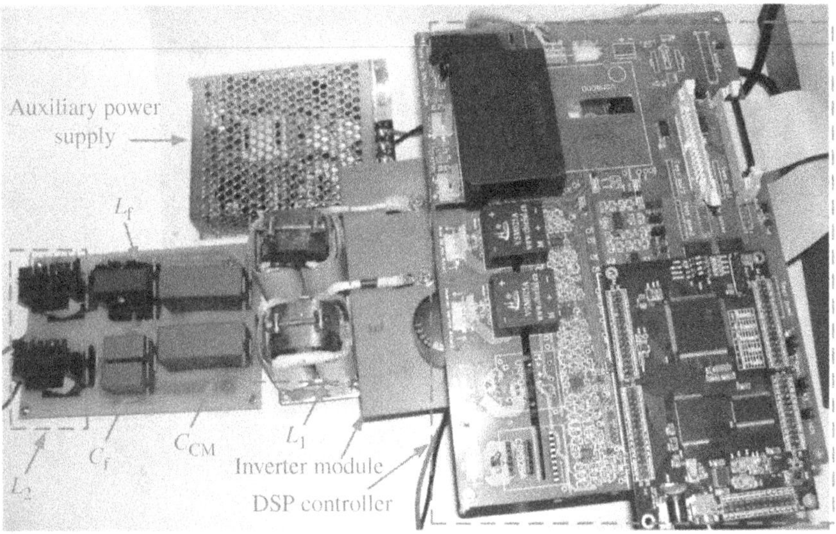

Figure 4.10 Photo of the modified *LLCL*-filtered GCI prototype.

4.3.4.1 Power Spectrum of the Grid-Injected Current

Figure 4.11 shows the measured power spectrum of grid-injected current for four different types of power filter-based GCIs.

Table 4.3 shows the measured attenuations of the maximum harmonic current for four different filter-based systems at approximately the switching frequency and twice the switching frequency. Compared to the conventional and modified *LCL* filters, the conventional and modified *LLCL* filters have better attenuating effects on the harmonics of the grid-side current at approximately the switching frequency; however, they have an opposite effect at approximately twice the switching frequency. Additionally, the modified *LLCL* filter has better attenuation on the harmonic than the conventional *LLCL* filter at approximately twice the switching frequency owing to the extra series split CM capacitors.

4.3.4.2 Measured Conducted EMI Noise

A spectrum analyzer (Agilent E4402) and LISN (EMCO 4825) were used to measure the EMI noise. In the spectrum analyzer, the peak value of the conducted EMI voltage was tracked. Figure 4.12 shows the measured background EMI noise when the auxiliary power supply of the controller is ON, and the output PWM signals are all blocked for four different types of systems. It can be seen that the background EMI noise is close to the standards of CSIPR 11 Class B at a frequency of approximately 150 kHz. However, no extra AC EMI filter was inserted during the test.

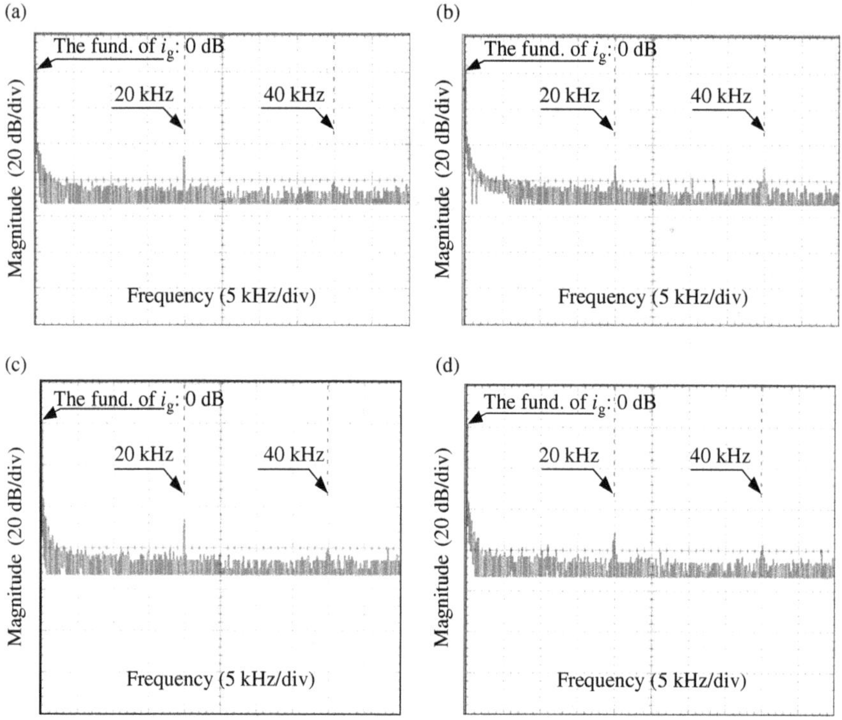

Figure 4.11 Measured power spectrum results of the grid-injected current for GCI with (a) conventional *LCL* filter; (b) conventional *LLCL* filter; (c) modified *LCL* filter; (d) modified *LLCL* filter.

Table 4.3 Attenuations of the maximum harmonic current for four different filter-based systems at approximately the switching frequency and twice the switching frequency.

Frequency	Conventional *LCL* filter (dB)	Conventional *LLCL* filter (dB)	Modified *LCL* filter (dB)	Modified *LLCL* filter (dB)
f_s	49	53	49	53
$2f_s$	62	54	62	57

Figure 4.13 shows the measured grid-side conducted EMI voltages of the four types of power filter-based systems. It can be seen that, for the conventional *LLCL* or *LCL* filter-based systems, the total conducted EMI noise cannot meet the standards of CSIPR 11 Class A. However, for the modified *LLCL* filter or *LCL* filter-based GCI, the total conducted EMI noise can meet the standards of CSIPR 11 Class A. Comparing the parameters of the modified *LLCL* filter with the

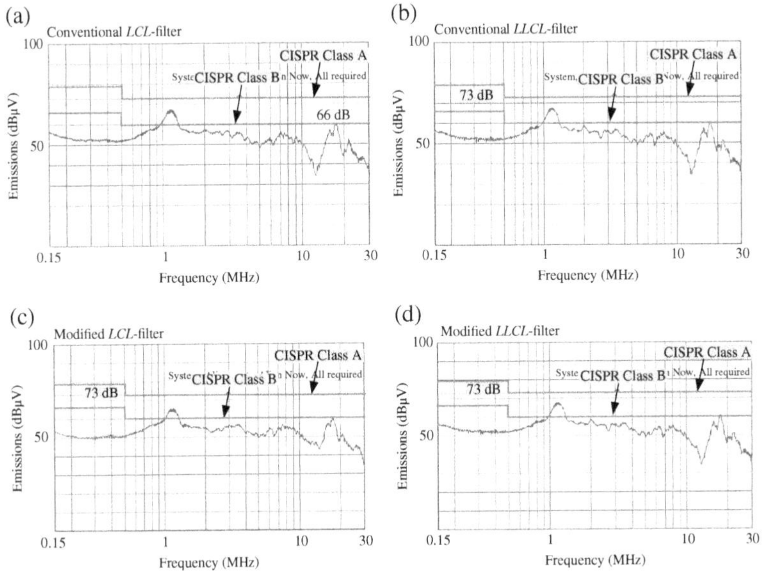

Figure 4.12 Measured background conducted EMI noise for different filters-based GCI (a) conventional *LCL* filter; (b) conventional *LLCL* filter; (c) modified *LCL* filter; (d) modified *LLCL* filter.

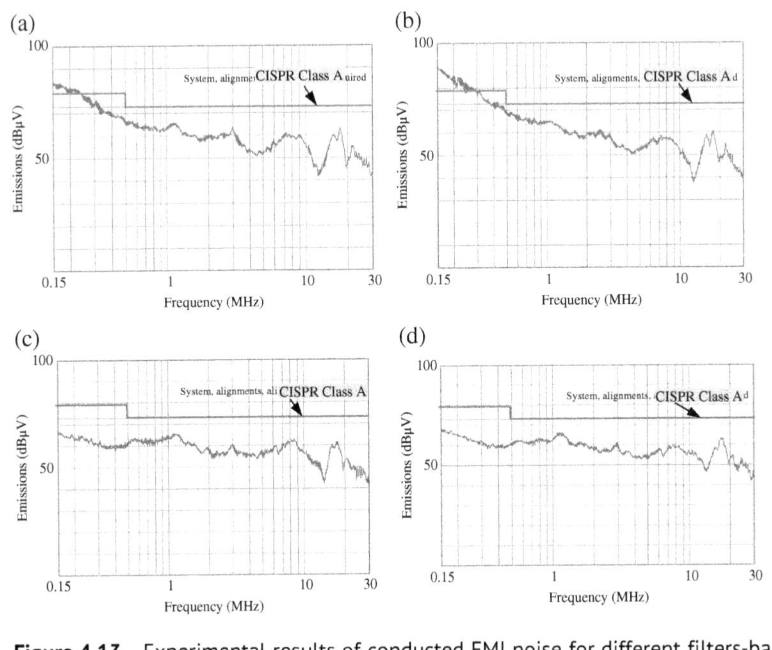

Figure 4.13 Experimental results of conducted EMI noise for different filters-based GCI (a) conventional *LCL* filter; (b) conventional *LLCL* filter; (c) modified *LCL* filter; (d) modified *LLCL* filter.

modified *LCL* filter as listed in Table 4.1, the total inductance of the *LLCL* filter is approximately 20% smaller than that of the *LCL* filter.

4.3.5 Negative DC-Rail Voltage with Respect to the Earth v_{dc_N} and Leakage Current

In PV inverter applications, the stray capacitance between the PV array and protection earth is generally proportional to the power rating of the inverter system. In the experimental prototype, two capacitors of 44 nF were used to emulate the stray capacitors C_p as shown in Figures 4.1 and 4.5. For the conventional *LLCL* filter and modified *LLCL* filter-based GCIs, the measured waveforms of the negative dc-rail voltage with respect to the protection earth v_{dc_N} and leakage current are shown in Figures 4.14 and 4.15, respectively. In the modified *LLCL* filter structure, the

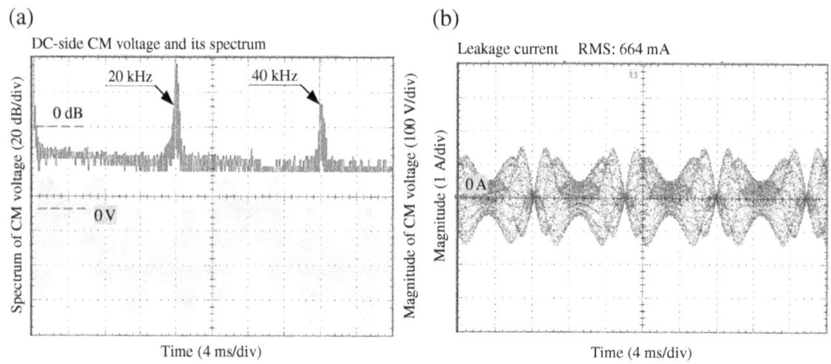

Figure 4.14 Measured waveforms of the negative dc-rail voltage with respect to the protection earth (v_{dc_N}) and leakage current for the conventional *LLCL*-filtered GCI (a) v_{dc_N}; (b) leakage current.

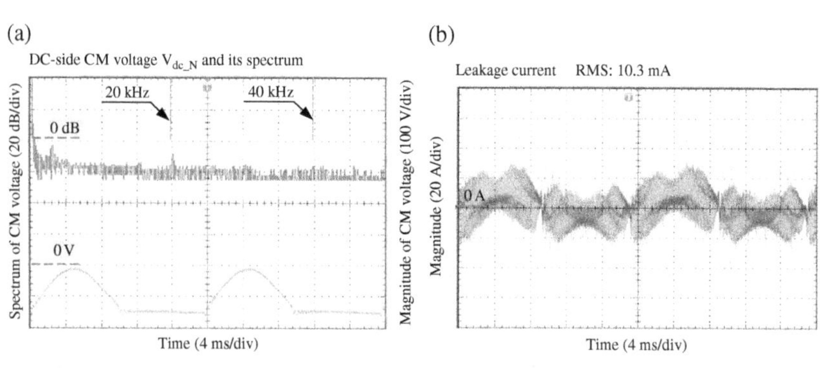

Figure 4.15 Measured waveforms of the negative dc-rail voltage with respect to the protection earth (v_{dc_N}) and leakage current for the modified *LLCL*-filtered GCI (a) v_{dc_N}; (b) leakage current.

negative dc-rail voltage of the system is smooth and RMS value of the leakage current can be attenuated from 664 to 10.3 mA, thus meeting the standards given in DIN V VDE V 0126-1-1 [12]. The leakage current waveform of the modified *LCL* filter-based GCI is similar to the modified *LLCL* filter and the measured RMS value of the leakage current is 9.4 mA. This value also meets the standards given in DIN V VDE V 0126-1-1.

4.4 Novel DM EMI Suppressor for *LLCL*-Filtered GCI without CM Noise Issue

In the previous section, we discussed the improved *LLCL* filter structure in detail for a single-phase transformerless GCI using DUPWM and two independent parallel capacitors to suppress both DM and CM EMI noise and reduce leakage currents. Nevertheless, connecting the midpoint of the DC voltage and bypass capacitor in the improved method results in a complicated structure and expanded footprint.

If BPWM is used for single-phase full-bridge inverters, the effects of CM current and CM EMI noise can be ignored. Another method to eliminate the adverse effect of $V_{CM}(t)$ is to use special modulations [13–17] or special topologies with an extra switch or switches on the AC or DC side to offer a freewheeling current loop and ensure that CM voltage is constant [18–29].

Therefore, in this section, we have focused on a novel type of DM EMI noise suppression that is an improvement of the *LLCL*-filtered GCI. This improved method can not only achieve the system's requirements for EMI noise but also minimize the additional reactive power.

4.4.1 Proposed DM EMI Suppressor

To measure the DM EMI noise, LISN is often connected between the inverter and grid as shown in Figure 4.16.

From the EMI analysis as shown in Figure 4.17, it can be deduced that a bypass capacitor can suppress the DM EMI noise for the *LLCL*-filtered GCI. In [30], a DM EMI suppressor of a capacitor C_{DM} for the *LLCL*-filtered GCI was proposed for three-phase applications. Its single-phase structure is shown in Figure 4.16 where $L_{DM} = 0\ \mu F$. The equivalent circuit used to analyze the DM EMI noise is shown in Figure 4.17 where $L_{DM} = 0\ \mu F$. The attenuation gains in the DM EMI voltage noise can be described using the following:

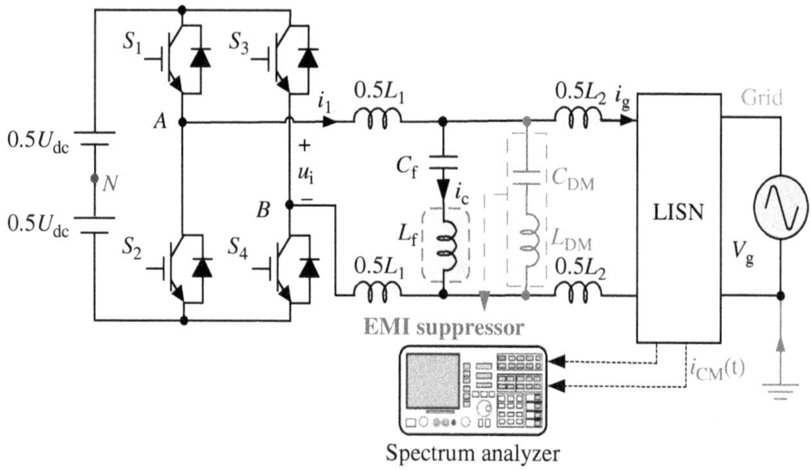

Figure 4.16 *LLCL*-filtered single-phase GCI connected to an LISN network.

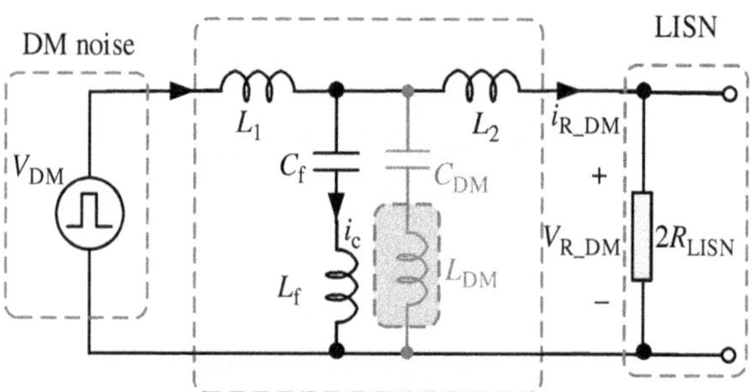

Figure 4.17 Equivalent circuit to analyze the DM EMI noise for *LLCL*-filtered system with parallel EMI suppressor.

$$\begin{aligned} &Att'_{DM\_LLCL\_1}(\omega)|dB|_{\omega \geq 2\pi \cdot 150kHz} \\ &\approx -20\log_{10}|Z_f//(j\omega L_2 + 2R_{LISN})//1/j\omega C_{DM} + j\omega L_1| \\ &\quad -20\log_{10}|Z_f//1/j\omega C_{DM} + j\omega L_2 + 2R_{LISN}| \\ &\quad + 20\log_{10}(2R_{LISN}) + 20\log_{10}|Z_f//1/j\omega C_{DM}| \end{aligned} \tag{4.16}$$

When $C_{DM} = 0.5\ \mu F$, the simulated DM EMI noise in the range of 150 kHz–1 MHz for *LLCL* filter-based system using the EMI suppressor of a capacitor is shown in Figure 4.18. As shown in Figure 4.18, the simulated results for the *LLCL* filter (with an EMI suppressor using a capacitor C_{DM})-based GCI match the requirement of CISPR Class A; however at a frequency of approximately

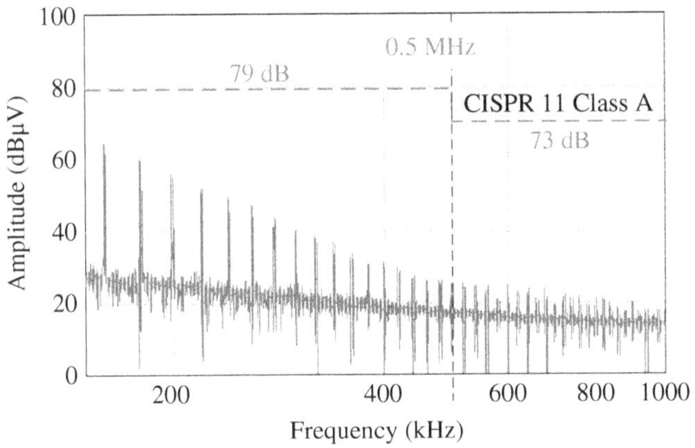

Figure 4.18 Simulated DM EMI noise in the range of 150 kHz–1 MHz with EMI suppressor of a capacitor (C_{DM} = 0.5 μF) for the *LLCL*-filtered GCI.

150 kHz, the DM EMI noise margin is at least 25 dB less than the *LCL*-filtered GCI as shown in Figure 4.4. If the capacitance of C_{DM} increases, the margin also improves. However, the additional reactive power also increases. To obtain sufficient noise attenuation at approximately 150 kHz, a L_{DM} C_{DM} series resonant circuit is proposed as a novel EMI suppressor, which is also shown in Figure 4.17. The attenuation gains on the EMI voltage noise can then be derived using the following:

$$\begin{aligned} &Att'_{\mathrm{DM}\_LLCL\_2}(\omega)|\mathrm{dB}|_{\omega \geq 2\pi \cdot 150\ \mathrm{kHz}} \\ &\approx -20\log_{10}|Z_f//(j\omega L_2 + 2R_{\mathrm{LISN}}) \times (1/j\omega C_{\mathrm{DM}} + j\omega L_{\mathrm{DM}}) + j\omega L_1| \\ &\quad -20\log_{10}|Z_f//(1/j\omega C_{\mathrm{DM}} + j\omega L_{\mathrm{DM}})| \\ &\quad + j\omega L_2 + 2R_{\mathrm{LISN}} + 20\log_{10}(2R_{\mathrm{LISN}}) \\ &\quad + 20\log_{10}|Z_f//(1/j\omega C_{\mathrm{DM}} + j\omega L_{\mathrm{DM}})| \end{aligned} \tag{4.17}$$

When C_{DM} = 0.5 μF and L_{DM} = 2.3 μH (the resonant frequency is set at 148.5 kHz), the simulated DM EMI noise in the range of 150 kHz–1 MHz is shown in Figure 4.19.

Comparing Figure 4.19 with Figure 4.18, it can be seen that at a frequency of approximately 150 kHz, the EMI noise attenuation of the proposed DM EMI suppressor is better than that of the EMI suppressor using only a capacitor. In the higher frequency range, despite the drawback of L_{DM}, the requirement of CISPR

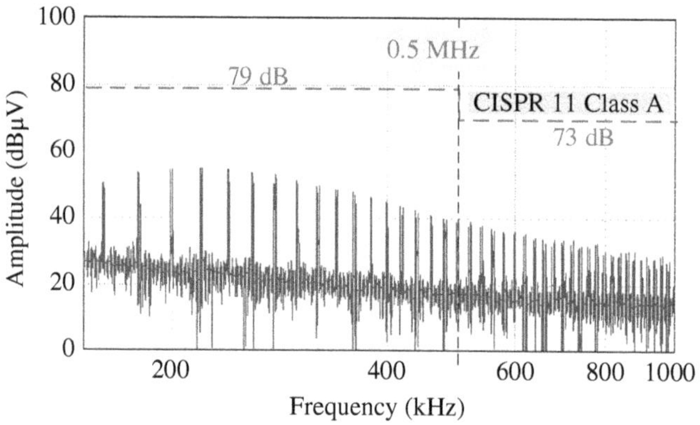

Figure 4.19 Simulated DM EMI noise in the range of 150 kHz–1 MHz with the proposed EMI suppressor of a series $L_{DM}C_{DM}$ resonant circuit for *LLCL*-filtered GCI.

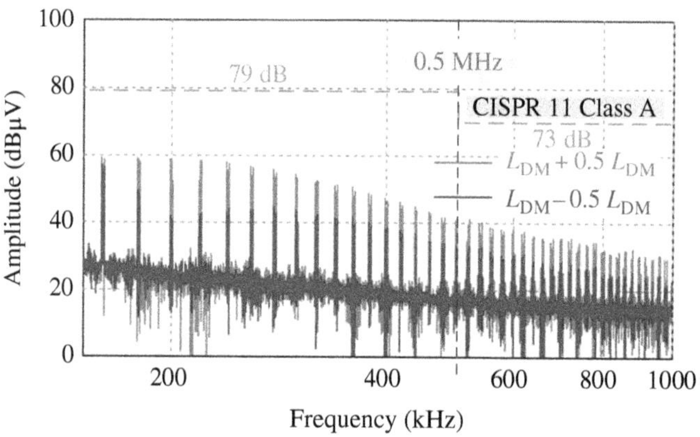

Figure 4.20 Sensitivity analysis of L_{DM} parameter drifting for the proposed DM EMI suppressor.

Class A can still be met for the *LLCL*-filtered GCI with the proposed DM EMI suppressor.

In real applications, parameter drifting is unavoidable especially for inductors with small values. Additionally, the EMI effect should be verified when the key parameter changes with a variation of 50%. Figure 4.20 shows the simulated EMI noise when the variation in L_{DM} is ±50% for *LLCL* filter-based system using the proposed EMI reduction method. It can be seen that the system is not very sensitive to the parameter drift of L_{DM}.

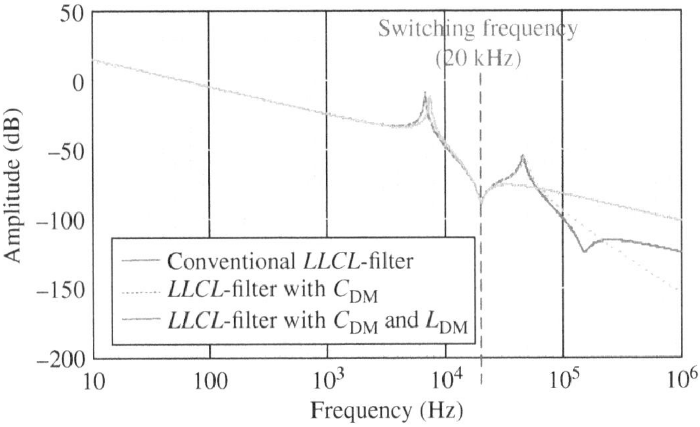

Figure 4.21 Amplitude plot of *LLCL* filter (grid-injected current vs. ac output voltage).

According to one of the previous studies [30], the characteristic frequency, which limits the controller design, partially depends on C_{DM}. It is also necessary to check and compare the frequency response analysis of systems with and without the EMI suppressor. Figure 4.21 shows the amplitude plot of *LLCL* filter (grid-injected current vs. ac output voltage). It can be seen that, in the switching frequency range (0–20 kHz), the characteristics of three types of *LLCL* filter are almost the same owing to the small value of C_{DM}. Thus, the extra DM EMI noise filter does not cause any extra control difficulty for the *LLCL*-filtered system.

4.4.2 Experimental Verification

To verify the theoretical analysis, an EMI testing was conducted on the conventional *LCL* filter, conventional *LLCL* filter, *LLCL* filter with the EMI suppressor of a capacitor (C_{DM} = 0.5 μF), and *LLCL* filter with the proposed EMI suppressor of L_{DM} C_{DM} resonant circuit (C_{DM} = 0.5 μF and L_{DM} = 2.3 μH)-based GCI. The main parameters of the *LCL* filter and *LLCL* filter are the same as those used for simulation as listed in Table 4.3.

A 500 W prototype of a single-phase full-bridge GCI with a DSP (TMS320L F2812A) controller was constructed. The experiments were evaluated and investigated under the given conditions of f_s = 20 kHz, U_{dc} = 175 V, U_g = 100 V/50 Hz, and P_{rated} = 500 W. Additionally, bipolar SPWM modulation method was adopted. A spectrum analyzer (Agilent E4402) and LISN (EMCO 4825) was used to measure the EMI noise while the peak value of the DM EMI voltage was tracked. ZFSCJ-2-2 of Mini Circuits was used as the power splitter. Because of the bandwidth of the

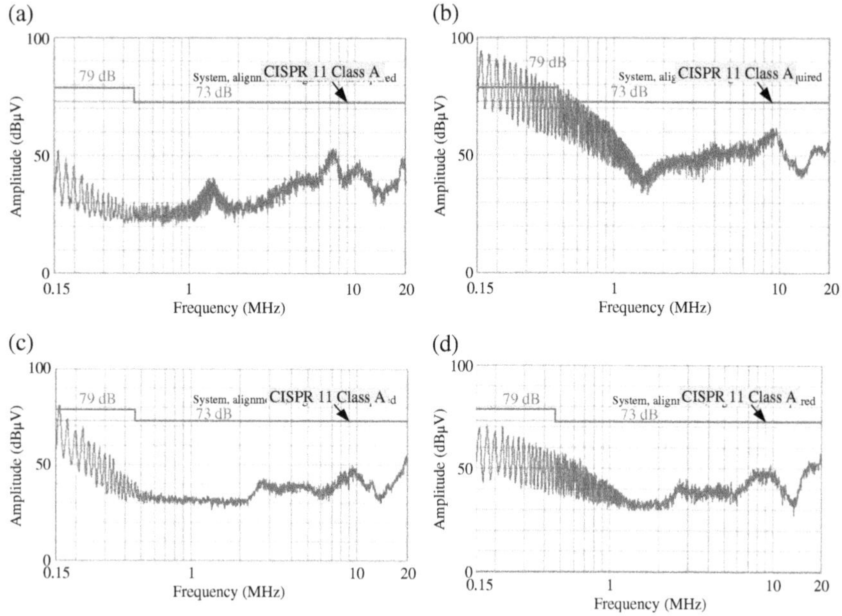

Figure 4.22 Measured DM EMI noise of (a) conventional *LCL*-filtered inverter system; (b) conventional *LLCL*-filtered inverter system; (c) *LLCL*-filtered inverter system with the EMI suppressor of C_{DM}; (d) modified *LLCL*-filtered inverter system with the proposed EMI suppressor of an $L_{DM}C_{DM}$ resonant circuit.

power splitter, the DM EMI noise only within the frequency range of 150 kHz–20 MHz can be measured precisely.

Figure 4.22 shows the measured DM EMI noise of the four different filter-based inverter systems. Figure 4.22a shows that if the grid inductor of the conventional *LCL* filter-based inverter system is designed to meet the injected current harmonics requirement of IEEE 1547.2-2008, then the EMI requirement of CISPR Class A with a large DM EMI noise margin can be met especially at a frequency of approximately 150 kHz. Figure 4.22b shows that the conventional *LLCL*-filtered inverter system cannot match the EMI requirement of CISPR 11 Class A. For the conventional *LLCL*-filtered inverter system, if an extra capacitor of 0.5 μF is connected in parallel to L_f C_f, the EMI requirement cannot be met as shown in Figure 4.22c. With an extra L_{DM} C_{DM} resonant circuit (C_{DM} = 0.5 μF and L_{DM} = 2.3 μH) in parallel with L_f C_f, the EMI requirement can be met with a margin of 9 dB μV as shown in Figure 4.22d.

Figure 4.22b–d indicate that, within the frequency range of 1–20 MHz, the measured EMI noises for the three types of *LLCL*-filtered systems can meet the requirement of CISPR 11 Class A quite well.

4.5 Summary

In this chapter, we combine two different applications to improve the ability of the single-phase transformerless GCI with *LCL*/*LLCL* filter to suppress the conducted EMI noise. The theoretical analysis was verified through simulations and experiments on a 0.5 kW/110 V/50 Hz single-phase GCI prototype, and the following conclusions can be drawn:

4.5.1 For Single-Phase Transformerless GCI Using DUPWM

1) For *LCL*-filtered system, the DM EMI noise met the requirement of CISPR 11 Class A with a large DM EMI noise margin at a frequency of approximately 150 kHz if the design of the grid-side inductor is based on the requirement of the harmonic current at approximately the switching frequency.
2) When extra EMI filters were not inserted, both the CM and DM EMI noises did not meet the requirements of CISPR 11 for the conventional *LLCL* filter-based system. Similarly, the CM EMI noises did not meet the requirements of CISPR 11 for the conventional *LCL*-filtered system.
3) Compared to the conventional *LCL*/*LLCL*-filtered inverter system, the attenuation of EMI noise of the modified *LCL*/*LLCL*-filtered system improved significantly, and thus the requirement of CISPR 11 Class A was met.
4) Compared to the conventional *LLCL* filter, the RMS value of leakage current of the modified *LLCL* filtered GCI decreased from 664 to 10.3 mA for a 500 W prototype with two stray capacitors of 44 nF, which meets the leakage current requirement of DIN V VDE V 0126-1-1.
5) Compared to the *LCL* filter, the modified *LLCL* filter reduced the total inductance by approximately 20%.

4.5.2 For Single-Phase Transformerless GCI Using BPWM or a System Without CM EMI Noise Issue

1) For the conventional *LLCL*-filtered GCI designed with normal parameters, the DM EMI noise did not meet the EMI requirement of CISPR 11 Class A in the frequency range of 150 kHz–1 MHz.

2) When the proposed DM EMI suppressor was adopted, the EMI requirement of CISPR 11 Class A was met for *LLCL*-filtered GCIs in the frequency range of 1–20 MHz.

The proposed EMI reduction method with $L_{DM}C_{DM}$ series resonant circuit performed better than the simple EMI reduction method with a capacitor of C_{DM} because it increased the attenuation of EMI noise at a frequency of approximately 150 kHz with the same additional reactive power.

References

1 Holmesand, D.G. and Lipo, T.A. (2003). *Pulse Width Modulation for Power Converters: Principles and Practice*. New York: Wiley-IEEE Press.

2 Kotny, J.-L., Margueron, X., and Idir, N. (2012). High-frequency model of the coupled inductors used in EMI filters. *IEEE Trans. Power Electron.* 27 (6): 2805–2812.

3 Tan, W., Cuellar, C., Margueron, X., and Idir, N. (2013). A high frequency equivalent circuit and parameter extraction procedure for common mode choke in the EMI filter. *IEEE Trans. Power Electron.* 28 (3): 1157–1166.

4 Wang, R., Blanchette, H.F., Mu, M. et al. (2013). Influence of high-frequency near-field coupling between magnetic components on EMI filter design. *IEEE Trans. Power Electron.* 28 (10): 4568–4579.

5 Sun, J. (2012). Conducted EMI modeling and mitigation for power converters and motor drives. *Proceedings of ESA Workshop Aerospace EMC*, pp. 1–6. Venice, Italy (21–23 May 2012). New York: IEEE.

6 Wu, W., He, Y., and Blaabjerg, F. (2012). An *LLCL*- power filter for single-phase grid-tied inverter. *IEEE Trans. Power Electron.* 27 (2): 782–789.

7 Wu, W., Sun, Y., Lin, Z. et al. (2014). A modified LLCL-filter with the reduced conducted EMI noise. *IEEE Trans. Power Electron.* 29 (7): 3393–3402.

8 Dong, D., Luo, F., Boroyevich, D., and Mattavelli, P. (2012). Leakage current reduction in a single-phase bidirectional AC–DC full-bridge inverter. *IEEE Trans. Power Electron.* 27 (10): 4281–4291.

9 Liserre, M., Blaabjerg, F., and Dell' Aquila, A. (2004). Step-by-step design procedure for a grid-connected three-phase PWM voltage source converter. *Int. J. Electron.* 91 (8): 445–460.

10 Liserre, M., Blaabjerg, F., and Hansen, S. (2005). Design and control of an *LCL* filter based three-phase active rectifier. *IEEE Trans. Ind. Appl.* 41 (5): 1281–1291.

11 Lopez, O., Teosorescu, R., Freijedo, F., and DovalGandoy, J. (2007). Leakage current evaluation of a single-phase transformer less PV inverter connected to the

grid. *Proceedings of the Applied Power Electronics Conference*, pp. 907–912. Anaheim, CA, USA (25 February–1 March 2007). New York: IEEE

12 Automatic Disconnection Device Between a Generator and the Public Low-Voltage Grid (2006). *DIN V VDE V 0126-1-1*. VDE Press.

13 Prasad, J.S.S., Ghosh, R., and Narayanan, G. (2015). Common-mode injection PWM for parallel converters. *IEEE Trans. Ind. Electron.* 62 (2): 789–794.

14 Lu, H., Cheng, X., Qu, W. et al. (2014). A three-phase current reconstruction technique using single dc current sensor based on TSPWM. *IEEE Trans. Power Electron.* 29 (3): 1542–1550.

15 Isomura, Y., Yamamoto, K., Morimoto, S. et al. (2014). Study of the further reduction of shaft voltage of brushless dc motor with insulated rotor driven by PWM inverter. *IEEE Trans. Ind. Appl.* 50 (6): 3738–3743.

16 Espina, J., Ortega, C., de Lillo, L. et al. (2014). Reduction of output common mode voltage using a novel SVM implementation in matrix converters for improved motor lifetime. *IEEE Trans. Ind. Electron.* 61 (11): 5903–5911.

17 Duran, M.J., Prieto, J., Barrero, F. et al. (2013). Space vector PWM with reduced common-mode voltage for five-phase induction motor drives. *IEEE Trans. Ind. Electron.* 60 (10): 4159–4168.

18 Buticchi, G., Barater, D., Lorenzani, E. et al. (2014). A nine-level grid-connected converter topology for single-phase transformerless PV systems. *IEEE Trans. Ind. Electron.* 61 (8): 3951–3960.

19 Matthias, V., Frank, G., Sven, B., and Uwe, H. (2006). H5-topology. *German Patent DE 102004030912 B3*.

20 Heribert, S., Christoph, S., and Juergen, K. (2003). HERIC-topology. *German Patent DE 10221592 A1*.

21 Gonzalez, R., Lopez, J., Sanchis, P., and Marroyo, L. (2007). Transformerless inverter for single-phase photovoltaic systems. *IEEE Trans. Power Electron.* 22 (2): 693–697.

22 Xiao, H.F., Liu, X.P., and Lan, K. (2014). Zero-voltage-transition full-bridge topologies for transformerless photovoltaic grid-connected inverter. *IEEE Trans. Ind. Electron.* 61 (10): 5393–5401.

23 Freddy Tan, K., Abd Rahim, N., Hew, W., and Che, H. (2015). Modulation techniques to reduce leakage current in three-phase transformerless H7 photovoltaic inverter. *IEEE Trans. Ind. Electron.* 62 (1): 322–331.

24 Kerekes, T., Teodorescu, R., Rodriguez, P. et al. (2011). A new high-efficiency single-phase transformerless PV inverter topology. *IEEE Trans. Ind. Electron.* 58 (1): 184–191.

25 Prasanna, U.R. and Rathore, A.K. (2014). Current-fed interleaved phase modulated single-phase unfolding inverter: analysis, design, experimental results. *IEEE Trans. Ind. Electron.* 61 (1): 310–319.

26 Zhang, L., Sun, K., Feng, L. et al. (2013). Family of neutral point clamped full-bridge topologies for transformerless photovoltaic grid-tied inverters. *IEEE Trans. Power Electron.* 28 (2): 730–739.

27 Ji, B., Wang, J., and Zhao, J. (2013). High-efficiency single-phase transformerless PV H6 inverter with hybrid modulation method. *IEEE Trans. Ind. Electron.* 60 (5): 2104–2115.

28 Bae, Y. and Kim, R. (2014). Suppression of common-mode voltage using a multicentral photovoltaic inverter topology with synchronized PWM. *IEEE Trans. Ind. Electron.* 61 (9): 4722–4733.

29 Barater, D., Buticchi, G., Lorenzani, E., and Concari, C. (2014). Active common mode filter for ground leakage current reduction in grid-connected PV converters operating with arbitrary power factor. *IEEE Trans. Ind. Electron.* 61 (8): 3940–3950.

30 Xu, J., Yang, J., Ye, J. et al. (2014). An LTCL filter for three phase grid-connected converters. *IEEE Trans. Power Electron.* 29 (8): 4322–4338.

Part III

Passive Damping

5

Design of Passive Damper for LCL/LLCL-Filtered Grid-Connected Inverter

5.1 Introduction

As illustrated in the previous chapters, the conventional *LCL* and *LLCL* filters can both be inserted between a voltage-source inverter (VSI) and grid to limit excess grid-injected current harmonics caused by sine-wave pulse-width modulation (PWM). However, regardless of the use of the *LCL* or *LLCL* filter, additional control difficulties must be considered, such as instability of the closed-loop control system and resonance between the inverter and power grid. To suppress any resonance caused by the high-order power filters, active damping [1–10] or passive damping [11–14] measures must be employed for stiff grid applications, the passive damping strategy is often more attractive owing to its simplicity and low cost. An effective passive damping method involves inserting an additional R_d–C_d circuit branch in the filter. When choosing the damping parameters of the high-order power filters, the harmonic attenuation, satisfactory damping effects, and additional power losses must be considered.

Compared with the *LCL* filter, the *LLCL* filter has an additional inductor, so the damping parameter design may be more complex and difficult to apply. Hence, an engineering design method for the passive damper of the high-order filter-based grid-connected inverter is desired, particularly for *LLCL* filter-based devices. This chapter presents the design of a passive damper for *LCL*/*LLCL*-filtered grid-connected inverters. First, the *LCL*/*LLCL* filter principles are presented for a single-phase grid-connected inverter. Then, a new engineering design method is introduced for optimal selection of the passive damping resistors by analyzing the impedance characteristics of the capacitor circuit and establishing the design criteria for passive damping *LCL* and *LLCL* filters. The power losses of the different damping methods are also compared and analyzed. Finally, comparative experiments are conducted with a 2 kW single-phase prototype device using a

Control and Filter Design of Single-Phase Grid-Connected Converters, First Edition. Weimin Wu, Frede Blaabjerg, Henry Chung, Yuanbin He, and Min Huang.

Published 2023 by John Wiley & Sons, Inc.

proportional resonant (*PR*) and harmonics compensation (*HC*) controller to verify the theoretical design. The results show that the proposed passive-damper-based *LLCL* filter not only reduces the total filter inductance and filter size but also decreases the damping power losses under stiff grid application conditions when compared with the *LCL* filter.

5.2 Design Method for Passive Damping

To attenuate any resonance caused by the high-order power filter irrespective of the use of the *LCL* or *LLCL* filter and to improve the stability of the closed-loop inverter system, passive or active damping schemes are adopted [15]. With regard to simplicity and cost, the passive damping method is more attractive than active damping. However, it is a challenge to balance the power losses, satisfactory damping effects, and harmonic attenuations in some cases when selecting the damping parameters for the high-order power filter [14]. Hence, this section presents a simple engineering damping design method.

5.2.1 Passive Damping Scheme of *LCL* Filter

The aim of damping is to reduce the *Q*-factor at the characteristic resonance frequency, which is often easily achieved by inserting a resistor in series with the capacitor, as shown in Figure 5.1a (called R_d damped *LCL* filter); its characteristic resonance frequency, ω_r, is calculated as:

$$\omega_r = \sqrt{\frac{L_1 + L_2}{L_1 L_2 C_f}} \tag{5.1}$$

Here, the inserted damping resistor does not affect the value of ω_r. To achieve appropriate damping, it is obvious that ω_r has to be limited within the shaded area in Figure 5.1b, which shows the impedance feature of the R_d–C_f circuit; herein, the

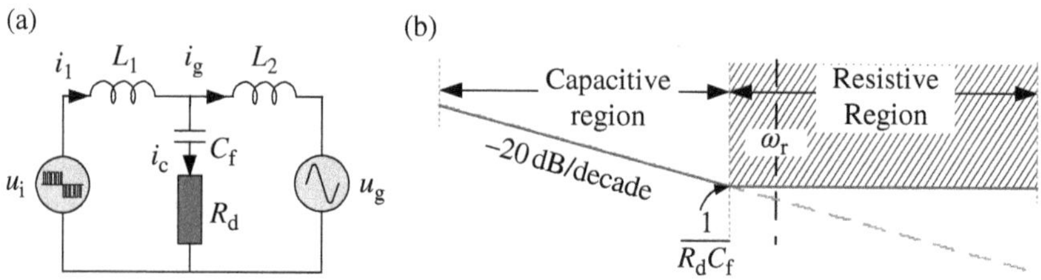

Figure 5.1 Characteristics of the passive damping method with R_d damper for the *LCL* filter. (a) schematic of the R_d damped *LCL* filter; (b) impedance feature of the capacitor branch (dotted line: without damping resistor).

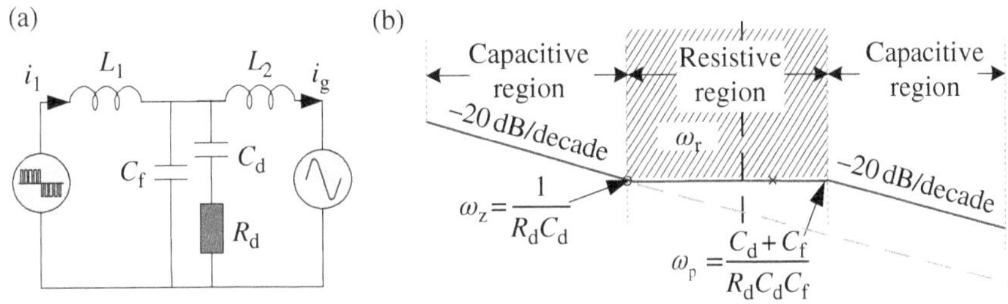

Figure 5.2 Characteristics of the improved passive damping method with R_d–C_d damper for the *LCL* filter. (a) Schematic of the R_d–C_d damped *LCL* filter; (b) impedance features of C_f and R_d–C_d damper (dotted line: without damping resistor).

capacitor branch represents the resistive feature for achieving a small *Q*-factor. This damping method is simple and effective, and the resistance R_d can be optimized by calculating the *Q*-factor of the filter [16, 17]. However, both the power loss and high-frequency harmonic attenuation are unsatisfactory with this method. To reduce the damping power loss and maintain high-frequency harmonic attenuation, an improved low-loss passive damping scheme [14] is introduced, as shown in Figure 5.2a (called R_d–C_d damped *LCL* filter).

It is assumed that the grid is an ideal voltage source, such that the transfer function $i_g(s)/u_i(s)$ can be calculated as follows:

$$G_{u_i \to i_g(\mathrm{LCL})}(s) = \left.\frac{i_g(s)}{u_i(s)}\right|_{u_g(s)=0} = \frac{R_d C_d s + 1}{s(L_1 L_2 R_d C_d C_f s^3 + L_1 L_2 (C_d + C_f) s^2 + R_d C_d (L_1 + L_2) s + L_1 + L_2)} \tag{5.2}$$

Compared with the former method, the advantage of this method is that the damping branch does not interfere with the high-frequency harmonic suppression of the *LCL* filter [7–10]. Unfortunately, unlike the former method, the characteristic resonance frequency of this approach varies for different R_d and C_d. For example, if $R_d = 0$, then,

$$\omega_{r\_down} = \sqrt{\frac{L_1 + L_2}{L_1 L_2 (C_f + C_d)}} \tag{5.3}$$

where ω_{r_down} is the minimum characteristic resonance frequency; if $R_d = \infty$, then,

$$\omega_{r\_up} = \sqrt{\frac{L_1 + L_2}{L_1 L_2 C_f}} \tag{5.4}$$

where ω_{r_up} is the maximum characteristic resonance frequency.

To determine the optimal Q-factor, the exact characteristic resonance frequency should be obtained by calculating the complex conjugate solutions of the third-order equation of the transfer function $i_g(s)/u_i(s)$ beforehand using the given R_d and C_d. Therefore, it is difficult to select reasonable damping parameters directly.

In this chapter, the design criteria of the *LCL* filter are almost similar to those presented in [18] except for the damping branch. Since the damping branch aims to decrease the Q-factor at the characteristic resonance frequency, a larger C_d/C_f is theoretically helpful for improving the damping effect but not the power losses. To balance the damping effects and damping losses, equal values of C_f and C_d may be used [14]. Moreover, the total capacitance must be limited to less than 5% of the apparent power at the rated load [17]. The impedance of C_f and the R_d–C_d damper, as shown in Figure 5.2a, $Z_{c(LCL)}(s)$, is calculated as:

$$Z_{c(LCL)}(s) = \frac{R_d C_d s + 1}{(R_d C_d C_f s + (C_d + C_f))s} \tag{5.5}$$

using the impedance features shown in Figure 5.2b.

For the same reason, to achieve good damping effect, the characteristic resonance frequency, $\omega_{r(LCL)}$, should be within the shaded area in Figure 5.2b, which is helpful for deriving the apparent resistance R_d. Accordingly, the resistance R_d is limited in the range:

$$\frac{1}{R_d C_d} \le \omega_{r(LCL)} \le \frac{C_d + C_f}{R_d C_d C_f} \tag{5.6}$$

Suppose the resistance R_d changes from 0 to ∞ Ω, it is easily deduced that:

$$\omega_{r\_down} \le \omega_{r(LCL)} \le \omega_{r\_up} \tag{5.7}$$

Combining (5.6) with (5.7), the rule for selecting the apparent resistance R_d is as follows:

$$\begin{cases} \dfrac{1}{R_d C_d} \le \omega_{r\_down} \\ \dfrac{C_d + C_f}{R_d C_d C_f} \ge \omega_{r\_up} \end{cases} \tag{5.8}$$

When $C_d/C_f = \lambda$, from (5.3), (5.4), and (5.8), the following equation can be derived:

$$\frac{\sqrt{\lambda + 1}}{\lambda}\sqrt{\frac{L_p}{C_f}} \le R_{d(LCL)} \le \frac{\lambda + 1}{\lambda}\sqrt{\frac{L_p}{C_f}} \tag{5.9}$$

where $L_p = L_1L_2/(L_1 + L_2)$ and R_d is localized within a narrow range. Furthermore, if the impedance of the grid varies over a wide range and the damping resistor cannot fully meet the condition in (5.8), a higher value of λ or L_2 may be needed.

5.2.2 Passive Damping Scheme of *LLCL* Filter

In Figure 5.3, two low-power-loss passive damping schemes are shown for the *LLCL* filter.

If the grid is an ideal voltage source, then the transfer function $i_g(s)/u_i(s)$ of the circuits in Figure 5.3a, b can be calculated as shown in (5.10) and (5.11), respectively,

$$G_{u_i \to i_g(\mathrm{LLCL\_1})}(s) = \left.\frac{i_g(s)}{u_i(s)}\right|_{u_g(s)=0} = \frac{s^3\alpha + s^2\beta + s\gamma + 1}{s^5a + s^4b + s^3c + s^2d + se} \tag{5.10}$$

where $a = L_1L_2L_fC_fC_d$, $b = (L_1L_2 + (L_1 + L_2)L_f)R_dC_dC_f$, $c = (L_1L_2(C_d + C_f) + L_fC_f(L_1 + L_2))$, $d = R_dC_d(L_1 + L_2)$, $e = (L_1 + L_2)$, $\alpha = L_fC_fC_dR_d$, $\beta = L_fC_f$, and $\gamma = R_dC_d$.

$$G'_{u_i \to i_g(\mathrm{LLCL\_2})}(s) = \left.\frac{i'_g(s)}{u'_i(s)}\right|_{u_g(s)=0} = \frac{s^3\alpha' + s^2\beta' + s\gamma' + 1}{s^4b' + s^3c' + s^2d' + se'} \tag{5.11}$$

where $b' = (L_1L_2 + (L_1 + L_2)L_f)R_dC_dC_f$, $c' = (L_1L_2(C_d + C_f) + L_f(C_f + C_d)(L_1 + L_2))$, $d' = R_dC_d(L_1 + L_2)$, $e' = (L_1 + L_2)$, $\alpha' = L_fC_f C_dR_d$, $\beta' = L_f(C_f + C_d)$, and $\gamma' = R_dC_d$.

From (5.10) and (5.11), it is noted that the characteristic resonance frequency cannot be easily calculated while selecting a reasonable damping resistor simultaneously, particularly for the case shown in Figure 5.3a. Comparing Figure 5.3b with Figure 5.2a, it is observed that the R_d–C_d damped *LLCL_2* filter differs from the R_d–C_d damped *LCL* filter in that the former has an extra inductor L_f. Moreover, L_f affects the characteristic resonance frequency boundaries as follows:

$$\omega'_{\mathrm{r\_down}} = \sqrt{\frac{1}{\left(\dfrac{L_1L_2}{L_1 + L_2} + L_f\right)(C_f + C_d)}} \tag{5.12}$$

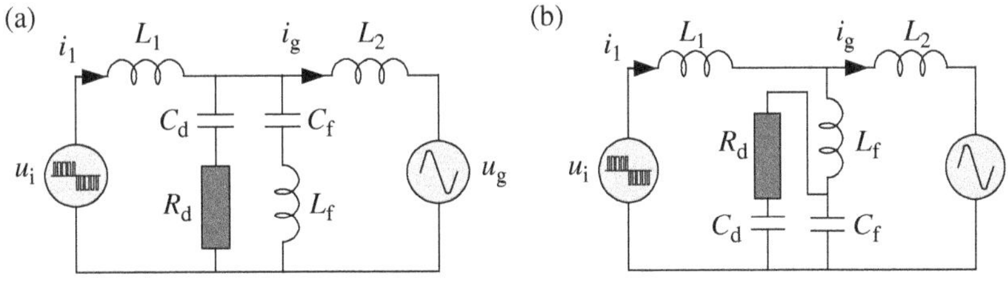

Figure 5.3 Proposed passively damped (a) *LLCL_1*; and (b) *LLCL_2* filters with R_d–C_d dampers.

$$\omega'_{\mathrm{r\_up}} = \sqrt{\frac{1}{\left(\frac{L_1 L_2}{L_1 + L_2} + L_{\mathrm{f}}\right) C_{\mathrm{f}}}} \tag{5.13}$$

Since the capacitive circuit in Figure 5.3b is the same as that in Figure 5.2a, the damping parameter selection for the R_{d}–C_{d} damped *LLCL_2* filter is similar to that of the R_{d}–C_{d} passive damped *LCL* filter. Further, to obtain a small *Q*-factor, the characteristic resonance frequency should be selected in the resistive region and should be lower than the switching frequency ω_{s}. The total capacitance $(C_{\mathrm{d}} + C_{\mathrm{f}})$ must also be limited to less than 5% of the apparent reactive power at the rated load. Similar to (5.9), the R_{d} of the R_{d}–C_{d} damped *LLCL_2* filter must meet the following requirement:

$$\frac{\sqrt{\lambda + 1}}{\lambda}\sqrt{\frac{L_{\mathrm{p}} + L_{\mathrm{f}}}{C_{\mathrm{f}}}} \leq R_{\mathrm{d(LLCL)}} \leq \frac{\lambda + 1}{\lambda}\sqrt{\frac{L_{\mathrm{p}} + L_{\mathrm{f}}}{C_{\mathrm{f}}}} \tag{5.14}$$

Figure 5.4 depicts the Bode plots of $i_{\mathrm{g}}(s)/u_{\mathrm{i}}(s)$ for the proposed damped *LLCL_1* and *LLCL_2* filters, which have almost similar characteristic resonance frequencies and frequency response characteristics within the switching frequency range such that C_{d} and R_{d} are very similar. Equation (5.14) can also be used to select C_{d} and R_{d} for the proposed passively damped *LLCL_1* filter. It is noted that the R_{d}–C_{d} damped *LLCL_2* filter has poorer harmonic attenuation than the *LLCL_1* filter and is not preferred for real applications.

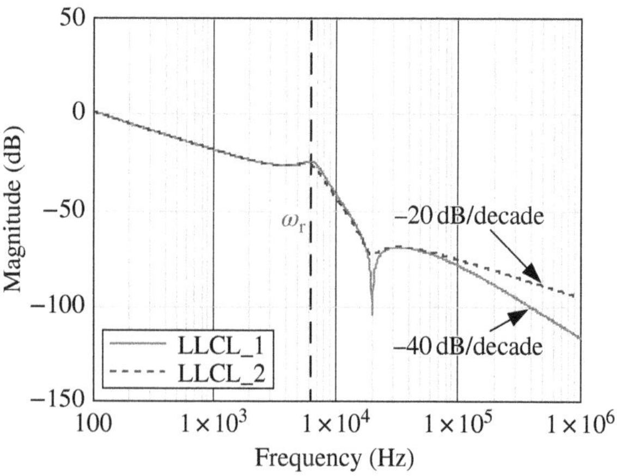

Figure 5.4 Bode plots of $i_{\mathrm{g}}(s)/u_{\mathrm{i}}(s)$ for the two R_{d}–C_{d} damped *LLCL* filters.

5.2.3 Design Example

Given that $f_s = 20$ kHz, $U_{dc} = 350$ V, $U_g = 220$ V/50 Hz, and $P_{rated} = 2$ kW and using discontinuous unipolar modulation [19], wherein two switches operate at the high frequency (20 kHz) and the other two operate at the line frequency (50 Hz), the filter parameters except for the damping parameters are designed according to the criteria in [18], which are listed in Table 5.1 (Case I is the passively damped *LCL* filter with R_d damper, Case II is the passively damped *LCL* filter with R_d–C_d damper, and Case III is the proposed passively damped *LLCL_1* filter); the resistors R_1, R_2, and R_f are the respective parasitic resistances of the inductors L_1, L_2, and L_f. It is noted that if continuous unipolar modulation is adopted, wherein the references of two phase legs are shifted by 180° and all switches are operated at the high frequency (20 kHz), the resonant frequency of L_f and C_f should be set to twice the switching frequency (i.e., 40 kHz).

The grid-side current ripple reductions at the switching frequency and twice the switching frequency are shown in Figure 5.5. It is seen that the passive damping method strongly inhibits resonance and improves system stability. Furthermore, Case I and Case III enable high-frequency harmonics attenuation by −40 dB/dec, as compared to the −60 dB/dec attenuation in Case II. It is seen that the designed *LLCL* filter has better attenuation effects on the harmonics than the *LCL* filter around the switching frequency, but this trend is reversed at twice the switching frequency. Thus, more studies on the resulting EMI issues are needed when using the *LLCL* filter-based system. The damping resistances for Case II and Case III in Table 5.1 are selected using (5.9) and (5.14). For example,

Table 5.1 Filter parameter values in the design.

Parameters	Case I R_d damped *LCL*	Case II R_d–C_d damped *LCL*	Case III R_d–C_d damped *LLCL*
L_1	1.2 mH	1.2 mH	1.2 mH
R_1	0.1 Ω	0.1 Ω	0.1 Ω
L_2	1.2 mH	1.2 mH	0.22 mH
R_2	0.04 Ω	0.04 Ω	0.01 Ω
L_f	—	—	32 μH
R_f	—	—	0.2 Ω
C_f	—	2 μF	2 μF
R_d	4 Ω	30 Ω	16.5 Ω
C_d	4 μF	2 μF	2 μF

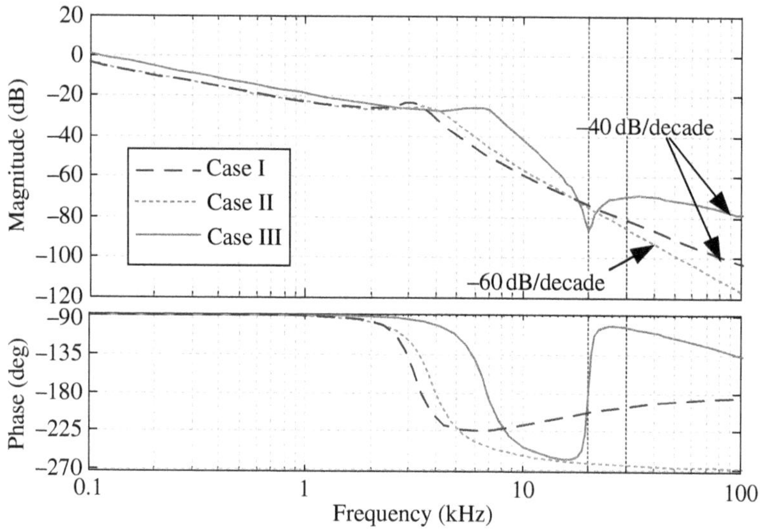

Figure 5.5 Bode plots of transfer functions from u_i to i_g for Case I (R_d damped *LCL* filter), Case II (R_d–C_d damped *LCL* filter), and Case III (R_d–C_d damped *LLCL_1* filter).

the ranges of the damping resistances for Case II and Case III are limited as $24.5\,\Omega \le R_{d(LCL)} \le 34.6\,\Omega$ and $14.8\,\Omega \le R_{d(LLCL)} \le 20.9\,\Omega$, respectively. Here, the damping resistances for Case II and Case III are selected as 30 and 16.5 Ω, respectively.

5.3 Analysis of Power Loss Caused by the Filter

The power losses resulting from the filter are attributed to the fundamental frequency current and high-frequency harmonic currents. Since the inverter-side inductor current ripples of both the *LCL* and *LLCL* filters are identical, the power losses of L_1 are also nearly similar. The difference in power loss is mainly caused by the remaining components, particularly L_2, L_f, and R_d.

5.3.1 Passive Damping Power Loss

When a single-phase full-bridge VSI is operated under asymmetric regularly sampled PWM conditions with discontinuous unipolar modulation, triangular carrier, and sinusoidal reference, its output voltage u_i can be derived as follows [19]:

$$u_i(t) = mU_{dc}\cos(\omega_o t) + \sum_{k=1}^{\infty}\sum_{n=\pm 1}^{\pm\infty} \frac{2U_{dc}J_n(k\pi m)}{k\pi}\sin\left(\frac{n\pi}{2}\right)\cos(k\omega_s t + n\omega_o t) \tag{5.15}$$

where m is the modulation index, U_{dc} is the DC-link voltage, and ω_o is the fundamental frequency in radians per second. Since the current of the grid-side inductor is expected to be a purely sinusoidal waveform, from the sideband harmonic currents of the inverter, the impedance can be approximated as:

$$Z_o(j\omega)\big|_{\omega\neq\omega_o} \cong L_1 j\omega \tag{5.16}$$

The harmonic amplitudes of the inverter-side current are given by:

$$|I_{AM}(n,k)|\big|_{\omega\neq\omega_o} = \frac{|U_{AM}(n,k)|}{|Z_o(j\omega)|} \tag{5.17}$$

where the amplitudes of the inverter output harmonic voltages $U_{AM}(n,k)$ are:

$$U_{AM}(n,k)\Bigg|_{\substack{k=1,2,\ldots,\infty \\ n=\pm1,\pm3,\ldots,\pm\infty}} = \left|\frac{2U_{dc}J_n(k\pi m)}{k\pi}\right| \tag{5.18}$$

As mentioned earlier, the power dissipated by the damping circuit is calculated in terms of the fundamental and harmonic currents separately. The final expressions of the damping losses are as follows:

1) Fundamental frequency damping loss: The fundamental frequency (50 or 60 Hz) damping losses of the filters in Figures 5.2a and 5.3a can be calculated approximately as:

$$P_{d\left(\frac{50}{60}\right)} = \frac{U_g^2 C_d^2 \omega_o^2 R_d}{1 + R_d^2 C_d^2 \omega_o^2} \tag{5.19}$$

2) High-frequency damping loss: Neglecting the deadtime effects and grid voltage harmonics, the dominant high-frequency current harmonics are composed of the switching frequency and twice the switching frequency. The high-frequency damping loss can be calculated as follows:

$$P_{d(\text{ripple})} = \left(\sum_{k=1,2}\ \sum_{n=\pm1,\pm3,\pm5} \frac{1}{2}\left(\frac{|Z_{LC}(j(n\omega_o+k\omega_s))||I_{AM}(n,\ k)|}{|Z_{RC}(j(n\omega_o+k\omega_s))+Z_{LC}(j(n\omega_o+k\omega_s))|}\right)^2\right)R_d \tag{5.20}$$

where $Z_{LC}(j(n\omega_o + k\omega_s))$ and $Z_{RC}(j(n\omega_o + k\omega_s))$ are the impedances of the L_f–C_f and R_d–C_d circuit branches, which can be expressed separately as:

$$Z_{LC}(j(n\omega_o + k\omega_s)) = L_f(j(n\omega_o + k\omega_s)) + \frac{1}{C_f(j(n\omega_o + k\omega_s))} \tag{5.21}$$

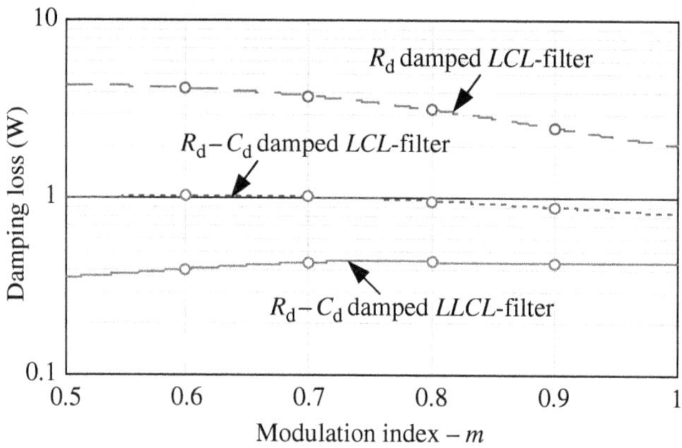

Figure 5.6 Damping loss comparison of the passively damped filters under different modulation indexes (at 2 kW power level).

$$Z_{\mathrm{RC}}(j(n\omega_o + k\omega_s)) = R_d + \frac{1}{C_d(j(n\omega_o + k\omega_s))} \tag{5.22}$$

The total damping loss is therefore given by:

$$P_{\mathrm{d(total)}} = P_{\mathrm{d}\left(\frac{50}{60}\right)} + P_{\mathrm{d(ripple)}} \tag{5.23}$$

The total damping power losses versus modulation index m for the different passively damped filters are shown in Figure 5.6. It is seen that the R_d damped LCL filter shown in Figure 5.1a has a larger damping loss, and the power loss of the R_d–C_d damped $LLCL$ filter shown in Figure 5.3a is smaller than that of the R_d–C_d damped LCL filter shown in Figure 5.2a since the former has a smaller damping resistor than the latter and the switching harmonic currents of the former are bypassed by the L_f–C_f circuit.

5.3.2 Power Losses in Inductors

The power loss in an inductor is made up of three components:

1) copper loss P_{cu};
2) iron loss P_{fe};
3) gap loss P_g.

In most cases, the skin and proximity effects contribute to the copper loss of the winding, so P_{cu} is equal to $I^2(R_{dc} + R_{ac})$, where the ac resistance R_{ac} is as specified in [20]. For the grid-side inductor L_2, the high-frequency ripple current is sufficiently attenuated by the filter to meet standard requirements such that the power

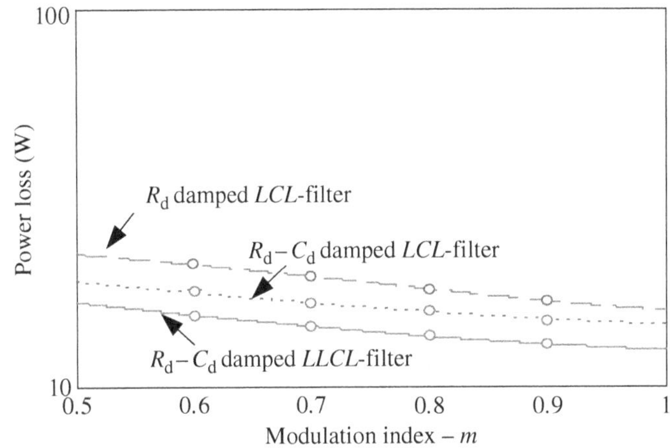

Figure 5.7 Power loss comparison of the different passively damped filters under different modulation indexes (at 2 kW power level).

loss is only due to the fundamental frequency current. Compared to the copper loss, the iron loss P_{fe} can be calculated with sufficient accuracy using the equations and data provided by the core manufacturers. The gap loss P_g is caused by the magnetic flux, which fringes around the gap and reenters the core, and is independent of the strip thickness and permeability of the core material. An accurate prediction of the gap loss depends on the amount of fringing flux [21] and is given by:

$$P_g = K_i E l_g f B_{ac}^2 \tag{5.24}$$

where K_i is the gap loss coefficient, E is the strip width of the core in centimeters, l_g is the gap length in centimeters, f is the operating frequency in hertz, and B_{ac} is the magnetic flux density in tesla.

The power losses of the different passively damped filters with respect to the modulation index m are shown in Figure 5.7. It is apparent that the R_d–C_d damped $LLCL$ filter has the lowest power loss.

5.4 Experimental Results

To compare the losses of the three different passive damping schemes and to verify the effectiveness of the R_d–C_d damped $LLCL$ filter, a 2 kW prototype based on a digital signal processor (DSP) (TMS320LF2407A) controller was constructed according to the circuit configuration shown in Figure 5.8. The parameters of

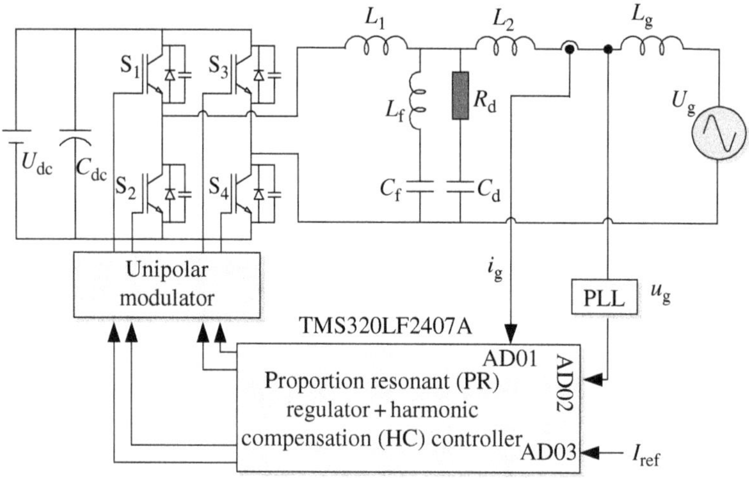

Figure 5.8 Prototype system configuration for testing different passive damping methods.

the filters are listed in Table 5.1, and the performance evaluation of the inverter with the R_d–C_d damped *LLCL* filter was evaluated under the conditions of $f_s = 20$ kHz, $U_{dc} = 350$ V, $U_g = 220$ V/50 Hz, $P_{rated} = 2$ kW, and deadtime $t_d = 2$ μs.

In Figure 5.8, the *PR* + *HC* (from 3rd to 9th) controller was selected for grid-current feedback control. The *PR* controller gain is expressed by (5.25), where K_p is the proportional gain and K_{ih} is the individual resonant integral gain:

$$G_{PR}(s) = K_p + \sum_{h=1,3,5,7,9} \frac{K_{ih}s}{s^2 + (\omega_o h)^2} \tag{5.25}$$

The magnitude of the proportional gain K_p from the *PR* controller determines the bandwidth and stability phase margin in the same manner as in the classical proportional–integral (PI) controller. The integral constant K_{ih} eliminates the steady-state error [15]. Figure 5.9 shows the Bode plots of the *PR* grid-side current control open-loop transfer function with $K_p = 0.22$ and $K_{ih} = 100$ for different grid impedances. Obviously, when the grid impedance increases from 0 to 3 mH, the system is able to maintain stability, but the bandwidth is reduced from 1500 to 570 Hz.

Referring to the analysis presented in Section III, an experimental analysis of the three cases is carried out for grid impedance $L_g = 0$ mH. Case I is the passively damped *LCL* filter with R_d damper, Case II is the passively damped *LCL* filter with R_d–C_d damper, and Case III is the R_d–C_d damped *LLCL* filter. The same *PR* + *HC* controller is used in all three cases. Figures 5.10–5.12 show the experimentally measured currents and spectra for the three cases. From these figures, the total

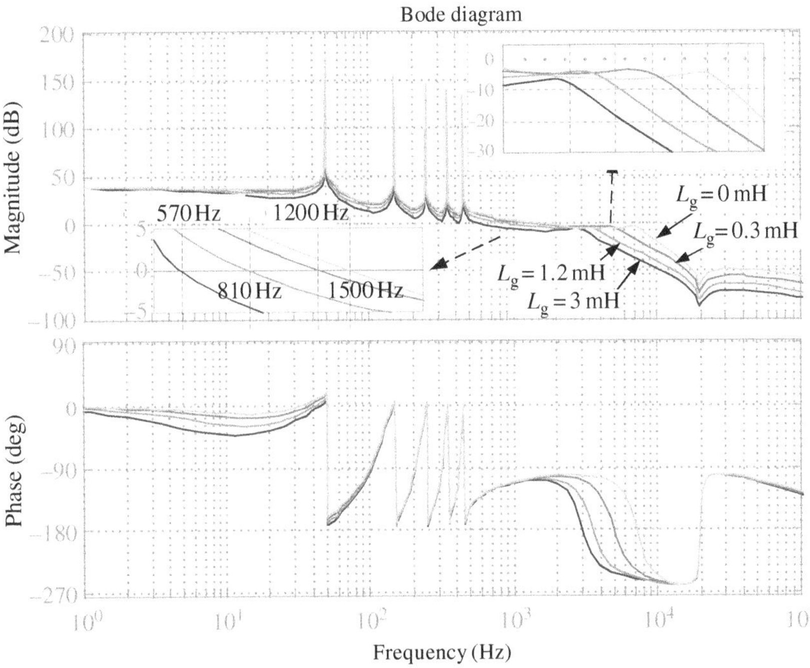

Figure 5.9 Bode plots of the open-loop transfer functions under different grid impedances.

harmonic distortion (THD) values of the grid-side currents in all three cases are nearly similar. Because of the wider control bandwidth, the low-frequency disturbance caused by the deadtime and crossover distortion is inhibited more effectively in Case III than Case I or Case II.

The spectra of the grid-side currents in the three cases are, respectively, shown in Figures 5.10b, 5.11b, and 5.12b, and the amplitudes of the dominant harmonic current (≥35th) are listed in Table 5.2. Figure 5.10b shows that the amplitude of the dominant harmonic current is −53 dB, which is about 0.22% of the fundamental (0 dB). Figure 5.11b shows that the amplitude of the dominant harmonic current is −51 dB, which is about 0.28% of the fundamental (0 dB). Figure 5.12b shows that the amplitude of the dominant harmonic current is −52 dB, which is about 0.25% of the fundamental (0 dB). Therefore, the amplitudes of the dominant harmonic grid-side currents (≥35th) in all cases meet the recommendation of the IEEE standard 519-1992, which states that each harmonic current exceeding the 35th should be less than 0.3% of the rated fundamental current of the grid-connected inverter system when the short-circuit current of the power system is below 20 times the nominal grid-side fundamental current.

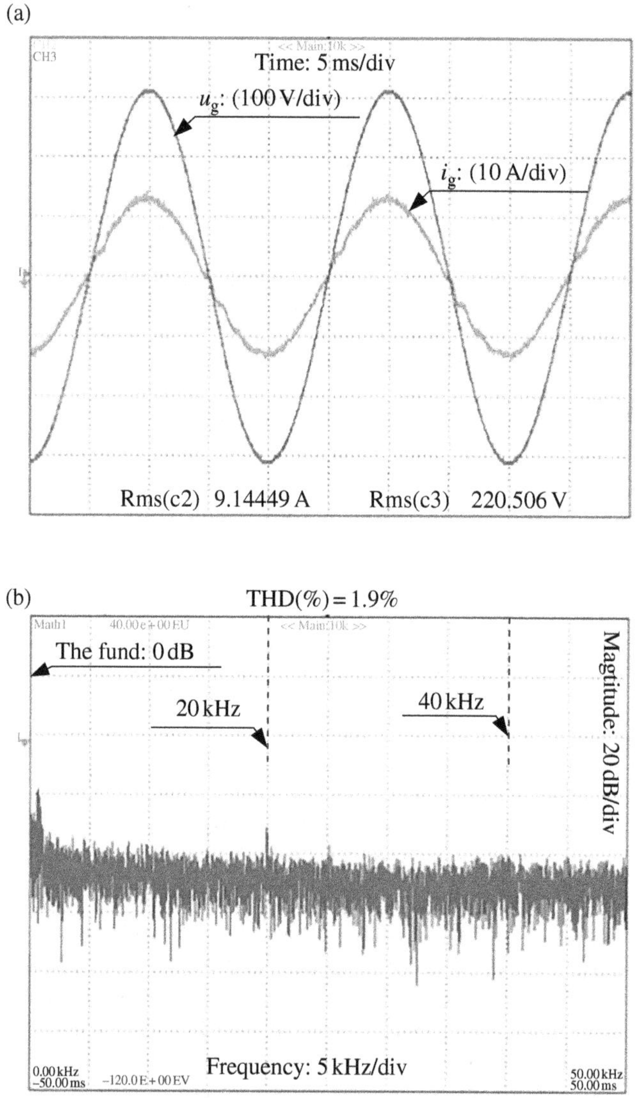

Figure 5.10 Experimental waveforms with R_d damped *LCL* filter (Case I with $I_g = 9.1$ A and $L_2 = 1.2$ mH). (a) Grid voltage and grid-side current waveforms; (b) power spectrum of the grid-side current.

The currents flowing through the damping branches for Case I, Case II, and Case III are shown in Figures 5.13–5.15, respectively, for a grid current of 9.1 A and grid voltage of 220 V. The measured and calculated damping losses are listed in Table 5.3.

(a)

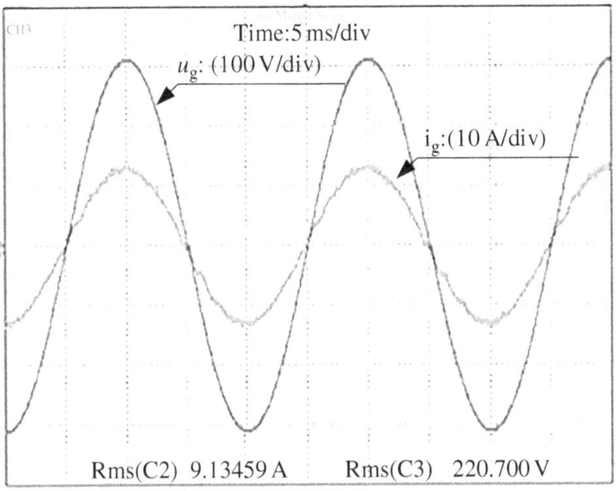

(b)

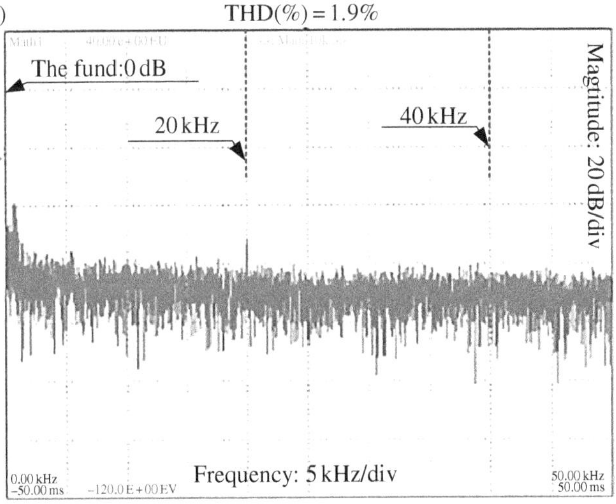

Figure 5.11 Experimental waveforms with R_d–C_d damped *LCL* filter (Case II with I_g = 9.1 A and L_2 = 1.2 mH). (a) Grid voltage and grid-side current waveforms; (b) power spectrum of the grid-side current.

Figures 5.16–5.18 show the experimental waveforms of the R_d–C_d damped *LLCL* filter-based single-phase grid-connected inverter for $C_d = 2\ \mu F$, $R_d = 16.5\ \Omega$, and grid impedances $L_g = 0.3$, 1.2, and 3 mH, respectively. It is seen that if the optimized damping resistor is set to the condition of zero grid impedance, increasing

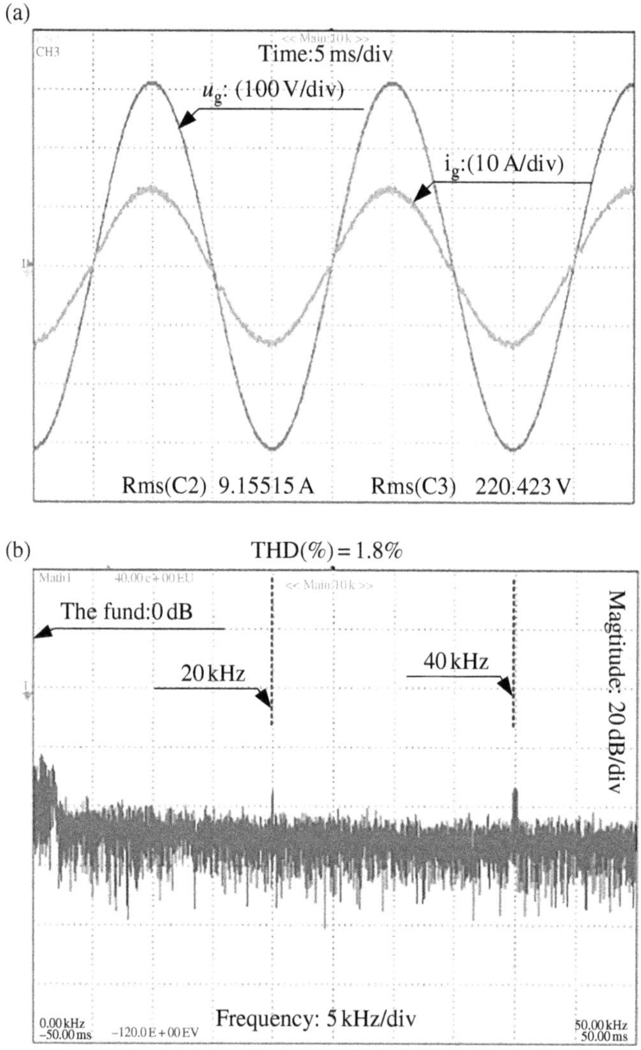

Figure 5.12 Experimental waveforms with R_d–C_d damped *LLCL* filter (Case III with I_g = 9.1 A and L_2 = 0.22 mH). (a) Grid voltage and grid-side current waveforms; (b) power spectrum of the grid-side current.

Table 5.2 Amplitudes of the dominant harmonic currents (≥35th).

Case I	Case II	Case III
0.22%	0.28%	0.25%

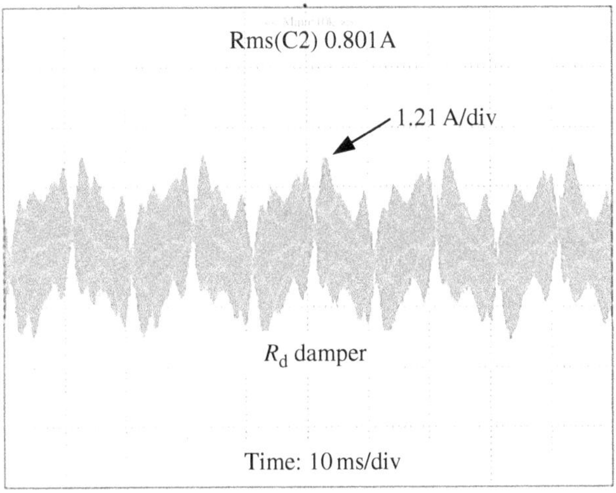

Figure 5.13 Current of R_d damper in Case I.

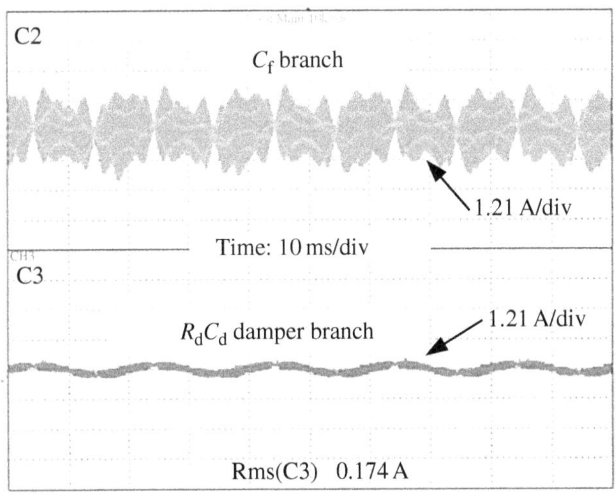

Figure 5.14 Currents of C_f and R_d–C_d damper branches in Case II.

the value of L_g will increase the distortion of the grid current, in accordance with the control analysis described in Figure 5.9. The distortion of a small grid current is worse than that of a large grid current, which is mainly due to the deadtime effects of the power converter.

Figure 5.19 shows the grid currents for an increased damping resistance of $R_d = 30\,\Omega$ for $L_g = 1.2$ and 3 mH. Comparing Figure 5.19 to Figures 5.17b and

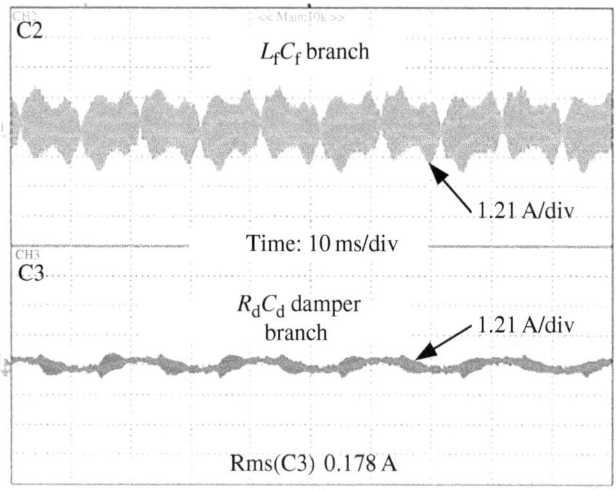

Figure 5.15 Currents of L_f–C_f and R_d–C_d damper branches in Case III.

Table 5.3 Damping losses in the three cases.

	Case I	Case II	Case III
Calculated losses (W)	2.51	0.89	0.45
Measured losses (W)	2.57	0.91	0.52

5.18b, it is seen that if the grid is weak, increasing the damping resistance R_d (along with a larger damping loss) is helpful for improving the THD of the grid current, which verifies the correction of (5.9) indirectly.

Furthermore, within half the switching frequency range that determines the low harmonics of the grid current, the *LLCL* filter-based grid-connected VSI has almost the same frequency response characteristics as the *LCL* filter, it is deduced that if only the low-frequency current harmonics are considered, the experimental currents of the R_d–C_d damped *LLCL* filter shown in Figure 5.19 are similar to those of the R_d–C_d damped *LCL* filter under the conditions of $L_2 = 1.2$ mH and $L_g = 0.22$ and 2.02 mH. The distortion of the grid current in Figure 5.19b is more serious than that in Figure 5.19a. Therefore, regardless of using the *LLCL* or *LCL* filter with simple passive damping, one cannot expect to achieve ideal grid current at any condition of the grid impedance.

For Case III, to verify the ability for attenuating current distortion resulting in background distortion from the grid voltage and to test the sensitivity to variations

(a)

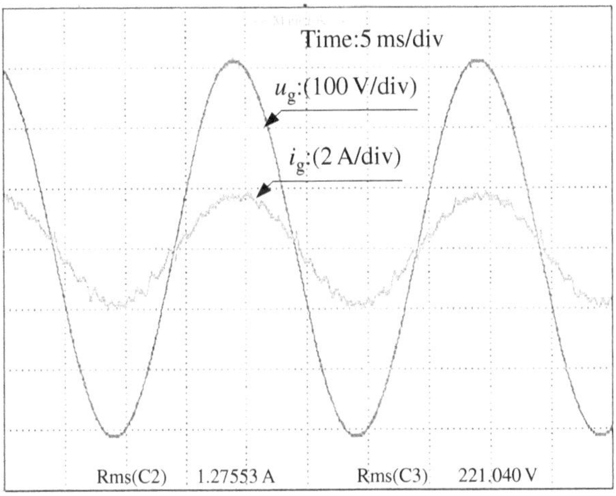

(b)

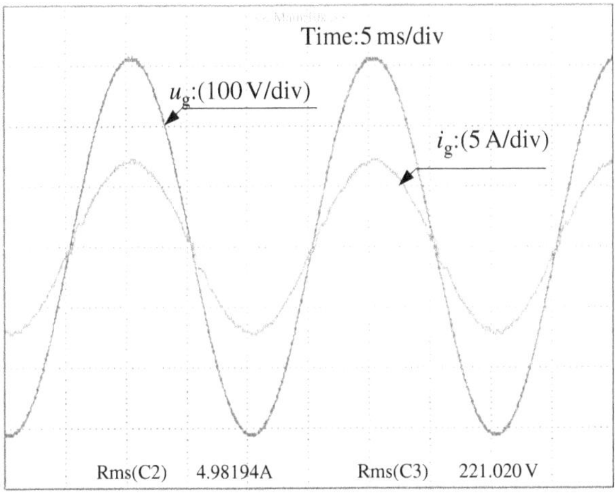

Figure 5.16 Experimental waveforms for R_d–C_d damped *LLCL* filter (L_g = 0.3 mH and R_d = 16.5 Ω). (a) Grid voltage and grid-side current (I_g = 1.3 A); (b) grid voltage and grid-side current (I_g = 5 A).

of the grid impedance, the 3rd, 5th, 7th, 9th, and 11th harmonics are injected into the grid voltage u_g, where the magnitudes of the injected harmonics with respect to the fundamental component of u_g are 1.1%, 2.8%, 1.4%, 2.3%, and 1.5% and corresponding phase angles are 180°, 0°, 180°, 0°, and 180°, respectively. The THD of the grid voltage is about 4.31%; here, the grid impedance is set to 0.8 mH, and the damping resistance is selected as 19 Ω. The measured 3rd, 5th, 7th, 9th, and 11th

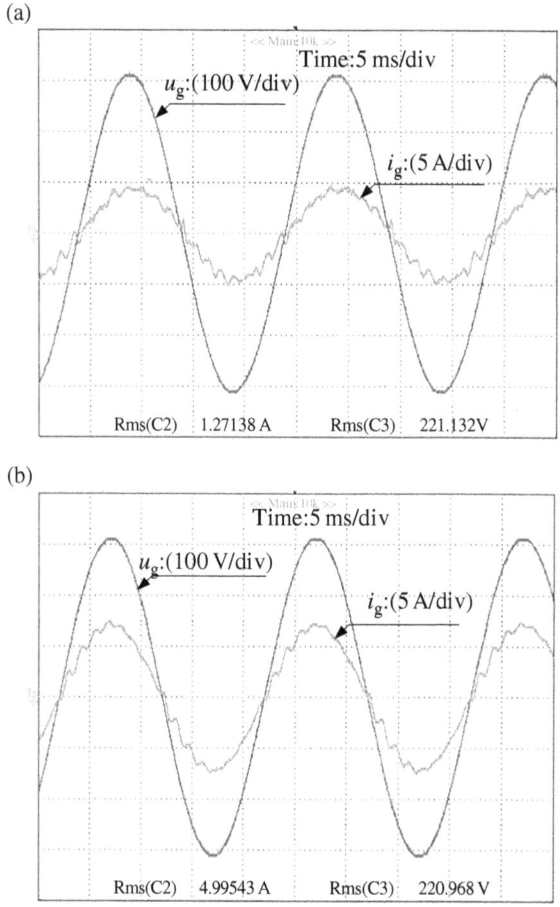

Figure 5.17 Experimental waveforms for R_d–C_d damped *LLCL* filter (L_g = 1.2 mH and R_d = 16.5 Ω). (a) Grid voltage and grid-side current (I_g = 1.3 A); (b) grid voltage and grid-side current (I_g = 5 A).

harmonic distortion values of the grid-side current waveforms shown in Figure 5.20 are 0.75%, 0.7%, 0.7%, 0.9%, and 0.75%, respectively. The output voltage of the inverter $U_i(t)$ is also shown. From Figure 5.20, it is seen that the grid-side current control system with the *PR* + *HC* controller has good ability to attenuate current distortions resulting from harmonics in the grid voltage.

5.5 Summary

This chapter presents a low-power-loss passively damped filtering scheme for the *LLCL* filter-based control system that was previously applied to the *LCL* filter-based grid-connected inverter; it uses the R_d–C_d series circuit in parallel with

(a)

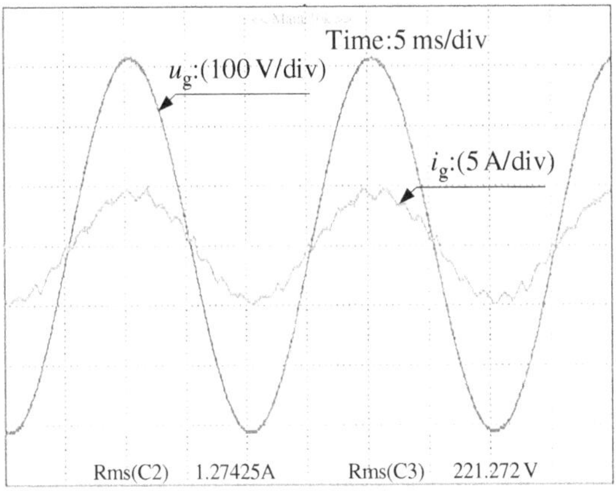

(b)

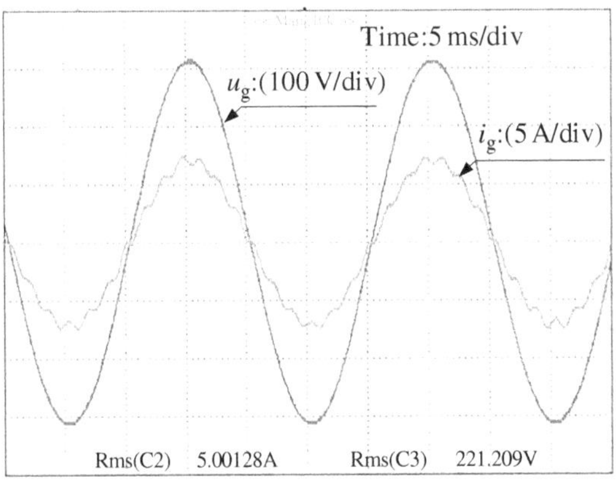

Figure 5.18 Experimental waveforms for R_d–C_d damped *LLCL* filter (L_g = 3 mH and R_d = 16.5 Ω). (a) Grid voltage and grid-side current (I_g = 1.3 A); (b) grid voltage and grid-side current (I_g = 5 A).

the capacitor of the *LCL* filter or L_f–C_f series resonant circuit of the *LLCL* filter to decrease the *Q*-factor of the characteristic resonance frequency. A simple engineering design method is presented through (5.9) and (5.14) for selecting the optimized damping resistor for the *LCL* filter and *LLCL* filter-based single-phase grid-connected inverter, respectively, while the capacitor C_d is determined by the desired maximum capacitive reactive power. This method is effective for not only the R_d–C_d damped *LLCL* filter-based system but also the R_d–C_d damped *LCL* filter-based system. Furthermore, the R_d–C_d damping method is more attractive for the *LLCL* filter-based system in stiff grid applications, where the damping power loss can be reduced since the harmonic currents around the switching frequency are mostly bypassed by the L_f–C_f series resonant circuit. A prototype 2 kW single-phase grid-connected VSI system with a *PR* + *HC* controller was

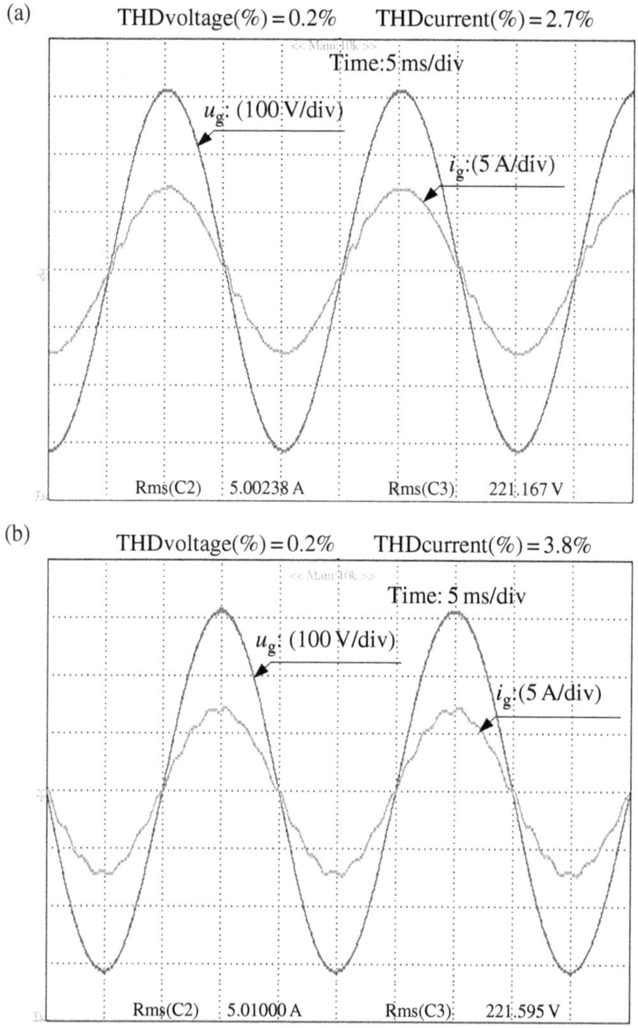

Figure 5.19 Experimental waveforms for R_d–C_d *LLCL* filter ($R_d = 30\ \Omega$). (a) Grid voltage and grid-side current ($L_g = 1.2$ mH and $I_g = 5$ A); (b) grid voltage and grid-side current ($L_g = 3$ mH and $I_g = 5$ A).

implemented to validate the R_d–C_d damped *LLCL* filter. The results show that if the grid is stiff, the system works well even though the THD of the grid voltage reaches 4.31%. Hence, research on EMI in the frequency range of 150 kHz–30 MHz using the *LLCL* filter is a natural next step in the overall assessment of the filter.

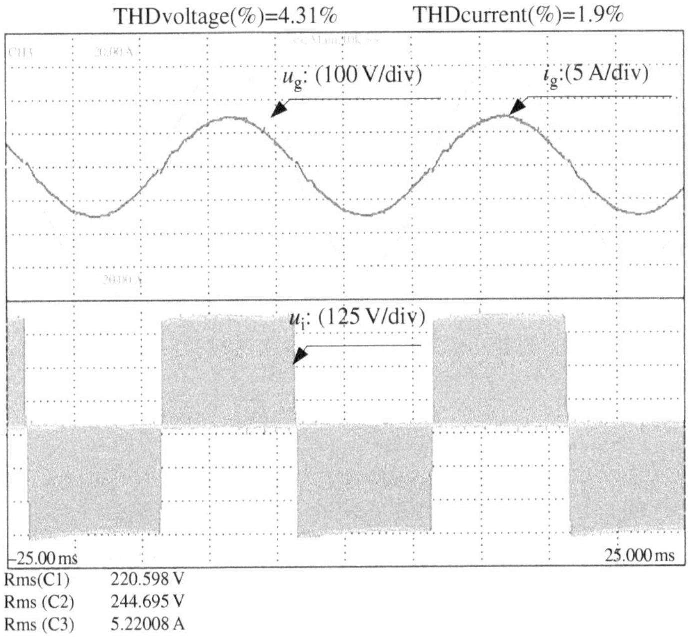

Figure 5.20 Output voltage of the inverter and grid-side current waveform with the *PR* + *HC* controller for Case III (I_g = 5 A, L_g = 0.8 mH, and R_d = 19 Ω) and with distortion in the grid.

References

1 Tang, Y., Loh, P., Wang, P. et al. (2012). Generalized design of high performance shunt active power filter with output LCL-filter. *IEEE Trans. Ind. Electron.* 59 (3): 1443–1452.

2 Tang, Y., Loh, P., Wang, P. et al. (2012). Exploring inherent damping characteristic of LCL-filters for three-phase grid-connected voltage source inverters. *IEEE Trans. Power Electron.* 27 (3): 1433–1443.

3 Dannehl, J., Liserre, M., and Fuchs, F.W. (2011). Filter-based active damping of voltage source converters with LCL filter. *IEEE Trans. Ind. Electron.* 58 (8): 3623–3633.

4 Cagnano, A., De Tuglie, E., Liserre, M., and Mastromauro, R.A. (2011). Online optimal reactive power control strategy of PV inverters. *IEEE Trans. Ind. Electron.* 58 (10): 4549–4558.

5 Shen, G., Zhu, X., Zhang, J., and Xu, D. (2010). A new feedback method for PR current control of LCL-filter-based grid-connected inverter. *IEEE Trans. Ind. Electron.* 57 (6): 2033–2041.

6 Espi, J.M., Castello, J., García-Gil, R. et al. (2011). An adaptive robust predictive current control for three-phase grid-connected inverters. *IEEE Trans. Ind. Electron.* 58 (8): 3537–3546.

7 Mohamed, Y.A.-R.I. (2011). Suppression of low- and high-frequency instabilities and grid-induced distortion and disturbances in distributed generation inverters. *IEEE Trans. Power Electron.* 26 (12): 3790–3803.

8 He, J. and Li, Y. (2012). Generalized closed-loop control (GCC) schemes with embedded virtual impedances for voltage source converters with LC or LCL filters. *IEEE Trans. Power Electron.* 27 (4): 1850–1861.

9 Massing, J.R., Stefanello, M., Grundling, H.A., and Pinheiro, H. (2012). Adaptive current control for grid-connected converters with LCL filter. *IEEE Trans. Ind. Electron.* 59 (12): 4681–4693.

10 Huerta, F., Pizarro, D., Cobreces, S. et al. (2012). LQG servo controller for the current control of LCL grid-connected voltage-source converters. *IEEE Trans. Ind. Electron.* 59 (11): 4272–4284.

11 Wang, T.C.Y., Ye, Z., Sinha, G., and Yuan, X. (2003). Output filter design for a grid-interconnected three-phase inverter. *Proceedings of the PESC*, pp. 779–784, Acapulco, NM (15–19 June 2003). New York: IEEE.

12 Turner, R., Walton, S., and Duke, R. (2010). Stability and bandwidth implications of digitally controlled grid-connected parallel inverters. *IEEE Trans.Power Electron.* 57 (11): 3685–3694.

13 Rockhill, A.A., Liserre, M., Teodorescu, R., and Rodriguez, P. (2011). Grid-filter design for a multimegawatt medium-voltage voltage-source inverter. *IEEE Trans. Ind. Electron.* 58 (4): 1205–1217.

14 Channegowda, P. and John, V. (2010). Filter optimization for grid interactive voltage source inverters. *IEEE Trans. Ind. Electron.* 57 (12): 4106–4114.

15 Liserre, M., Teodorescu, R., and Blaabjerg, F. (2006). Stability of photovoltaic and wind turbine grid-connected inverters for a large set of grid impedance values. *IEEE Trans. Power Electron.* 21 (1): 263–272.

16 Cha, H. and Vu, T.K. (2010). Comparative analysis of low-pass output filter for single-phase grid-connected photovoltaic inverter. *Proceedings of the IEEE APEC*, pp. 1659–1665, Palm Springs, CA (21–25 February 2010). New York: IEEE.

17 Liserre, M., Blaabjerg, F., and Hansen, S. (2005). Design and control of an LCLfilter-based three-phase active rectifier. *IEEE Trans. Ind. Appl.* 41 (5): 1281–1291.

18 Wu, W., He, Y., and Blaabjerg, F. (2012). An LLCL power filter for single-phase grid-tied inverter. *IEEE Trans. Power Electron.* 27 (2): 782–789.

19 Holmes, D.G. and Lipo, T.A. (2003). *Pulse Width Modulation for Power Converters: Principles and Practice*, 183–199. Hoboken, NJ: Wiley-IEEE Press.

20 Bartoli, M., Reatti, A., and Kazimierczuk, M.K. (1995). Modeling winding losses in high frequency power inductors. *J. Circuits, Syst. Comput.* 5 (4): 607–626.

21 Wm, C. and McLyman, T. (2004). *Transformer and Inductor Design Handbook*, 3e, 295–307. Boca Raton, FL: CRC Press.

6

Composite Passive Damping Scheme for *LLCL*-Filtered Grid-Connected Inverter

6.1 Introduction

In the previous chapter, we discussed the design of passive dampers for *LCL*/*LLCL* filtered grid-connected inverters. In this chapter, we focus on the composite passive damping scheme. When PWM is applied, the switching frequency is very large and results in harmonic generation. Therefore, use of the PWM scheme requires an output filter to limit the harmonic content of the grid-injected current. Here, an *LCL* filter is more acceptable than an *L* filter owing to its smaller size, lower cost, and better dynamics [1], particularly for applications exceeding several kilowatts. To further reduce the inductor size, a novel high-order power filter (*LLCL* filter) was proposed previously [2].

The background harmonics (harmonics before filtering) of the grid voltage result in distortions of the injected grid current of the grid-connected inverter. To minimize the effects of the grid background harmonic voltages (e.g. third to ninth), the proportional resonant and harmonic compensation ($PR + HC$) controller has become a popular approach in grid-connected inverters. Using the $PR + HC$ controller, the system open-loop gain, control bandwidth, and system stability margin are determined from the proportional gain K_p of the controller [3]. Moreover, large variations of the grid inductance affect the stability of the high-order power-filter-based system [4].

It is known that a passive damper can reduce the Q-factor at the dominant resonant frequency. For a high-order filter-based system, the optimized Q-factor is difficult to obtain by direct analysis; however, we can replace the original complex circuit with a relatively simple equivalent circuit without changing the nature of the original circuit. This method is referred to as the E-Q-factor analysis and is used to calculate the Q-factor. The passive damping method improves the stability of the

Control and Filter Design of Single-Phase Grid-Connected Converters, First Edition. Weimin Wu, Frede Blaabjerg, Henry Chung, Yuanbin He, and Min Huang.

Published 2023 by John Wiley & Sons, Inc.

system. An RC parallel and RL series dampers are often adopted for this purpose [5]. The RC parallel damping method is suitable for grids with relatively narrow ranges of the inductance values, whereas RL series damping method is only suitable for stiff grid conditions. To improve the system stability, a composite damping method is proposed for the *LLCL* filter, which effectively suppresses resonance for inductance variations over a wide range.

A discussion is presented based on the results and observations from the previous chapter. In Section 6.2, both the upper and lower limits of the *PR* + *HC* controller gain are analyzed first. In Section 6.3, we introduce the equivalent *Q*-factor (*E*-*Q*-factor) calculation method to select the optimal passive damping parameters. In Section 6.4, a composite damping method is proposed for the *LLCL* filter to overcome the adverse effects of large variations of the grid inductance and to suppress any resonance. In Section 6.5, simulations and experiments are presented for a 2 kW single-phase grid-connected inverter prototype to confirm the accuracy of the theoretical analysis. Lastly, the conclusions of the above methods are summarized in Section 6.6.

6.2 Upper and Lower Limits of the *PR* + *HC* Controller Gain

When the *PR* + *HC* controller is adopted, the system open-loop gain, control bandwidth, and system stability margin are determined from the proportional gain, K_p, of the controller [3]. In this section, the upper and lower limits of K_p are analyzed.

6.2.1 *LLCL* Filter-Based Grid-Connected Inverter Configuration

Figure 6.1 shows a typical grid-connected inverter with an *LLCL* filter.

The transfer function of the grid-injected current versus the output voltage of the inverter is derived as follows:

$$G_{u_i \to i_g}(s) = \left.\frac{i_g(s)}{u_i(s)}\right|_{u_g(s)=0} = \frac{L_f C_f s^2 + 1}{(L_1(L_2+L_3))C_f + (L_1 + (L_2 + L_g))L_f C_f s^3 + (L_1 + (L_2 + L_g))s} \tag{6.1}$$

If L_f is set to zero, then the transfer function $i_g(s)/u_i(s)$ of the classical *LCL* filter can also be derived. Figure 6.2 shows the diagram of the system with grid-side current feedback control, where $G_c(s)$ denotes the *PR* + *HC* controller, $H(s)$ is

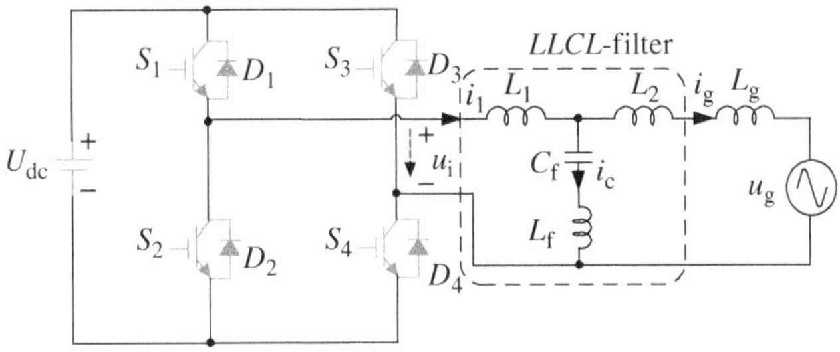

Figure 6.1 VSI connected to the grid through the *LLCL* filter.

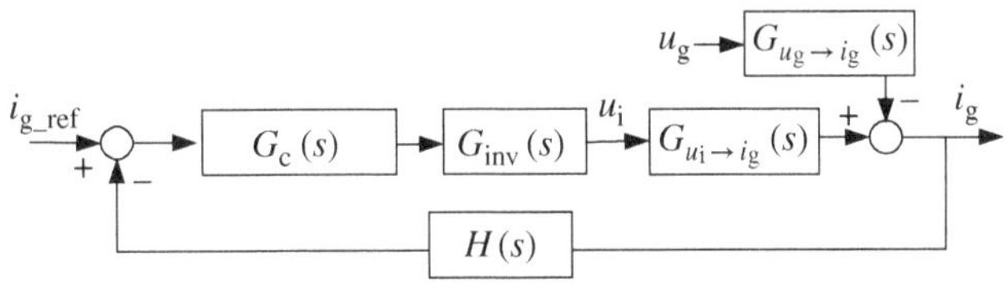

Figure 6.2 Control block of the system using the grid-side current feedback.

the sensor gain of the grid-injected current, and $G_{inv}(s)$ is the gain of the PWM inverter.

The open-loop transfer function of the system and magnitude of the open-loop gain are as follows:

$$G_{\text{open-loop}}(s) = G_c(s)G_{u_1 \to i_g}(s)G_{inv}(s)H(s) \tag{6.2}$$

$$\left|G_{\text{open-loop}}(j\omega)\right| = 20\log\left|G_c(j\omega)G_{u_i \to i_g}(j\omega)G_{inv}(j\omega)H(j\omega)\right| \tag{6.3}$$

6.2.2 Lower Limit of the *PR* + *HC* Controller Gain

To include the *PR* + *HC* controller in the system, the crossover frequency should be greater than or equal to a set value of f_{min_cross} under the weakest grid condition. In this work, f_{min_cross} is designated as 500 Hz to suppress up to the ninth harmonic current. In the low-frequency range, the *LCL* or *LLCL* filter has approximately the same frequency characteristics as the *L* filter, and the loop gain is unity at the crossover frequency, so the lower limit of the *PR* + *HC* controller gain can be derived as follows:

$$k_{p\_min} = \frac{2\pi f_{min\_cross}(L_1 + L_2 + L_g)}{\left|G_{inv}(j2\pi f_{min\_cross})H(j2\pi f_{min\_cross})\right|} \tag{6.4}$$

6.2.3 Upper Limit of the *PR* + *HC* Controller Gain

The derivation of the upper limit for the controller gain can be categorized under two steps. First, a conservative upper limit, $k_{p\_max\_1}$, is derived on the basis of the gain margin of the system open-loop transfer function. Then, considering the acceptable phase margin, the second upper limit, $k_{p\_max\_2}$, is obtained. Thus, the designed controller gain is characterized by

$$k_{p\_min} \leq k_p < \min\left(k_{p\_max\_1}, k_{p\_max\_2}\right) \tag{6.5}$$

Step 1: Controller gain based on the gain margin – for a conservative controller design under different grid conditions, the open-loop gain $|G_{\text{open-loop}}(j\omega)|$ should be less than 0 dB whenever ω exceeds the crossover frequency (radians per second). Assuming that the transfer function of the grid-side current versus the output voltage of the inverter with passively damped high-order filter is $G'_{u_i \to i_g}(s, L_g)$, the gain of the controlled plant $\left|G'_{\text{plant}}(j\omega)\right|$ is given by

$$\left|G''_{\text{plant}}(j\omega)\right| = 20 \log \left|G_{u_1 \to i_g}(j\omega) G_{\text{inv}}(j\omega) H(j\omega)\right| \tag{6.6}$$

Once the parameters of the passively damped high-order filter are chosen, $\left|G''_{\text{plant}}(j\omega)\right|$ is a function of ω and L_g; its first-order partial derivative and Hessian matrix can be derived as shown in (6.7) and (6.8), respectively:

$$\begin{aligned} \frac{\partial\left|G''_{\text{plant}}(j\omega, L_g)\right|}{\partial \omega} &= 0 \\ \frac{\partial\left|G''_{\text{plant}}(j\omega, L_g)\right|}{\partial L_g} &= 0 \end{aligned} \tag{6.7}$$

$$H_f = \begin{vmatrix} \dfrac{\partial^2\left|G'_{\text{plant}}(j\omega, L_g)\right|}{\partial \omega^2} & \dfrac{\partial^2\left|G'_{\text{plant}}(j\omega, L_g)\right|}{\partial L_g \partial \omega} \\ \dfrac{\partial^2\left|G'_{\text{plant}}(j\omega, L_g)\right|}{\partial \omega \partial L_g} & \dfrac{\partial^2\left|G'_{\text{plant}}(j\omega, L_g)\right|}{\partial L_g^2} \end{vmatrix} \tag{6.8}$$

From (6.7) and (6.8), the frequency $\omega_{\max\_1}$ and grid inductance $L_{g\_max\_1}$ corresponding to the local maximum gain of the plant can be calculated. Since the *PR* + *HC* controller is adopted, the conserved gain of the controller can be derived as

$$K_{p\_max\_1} = \frac{1}{\left|G''_{u_i \to i_g}(j\omega_{\max\_1}, L_{g\_max\_1}) G_{\text{inv}}(j\omega_{\max\_1}) H(j\omega_{\max\_1})\right|} \tag{6.9}$$

Step 2: Controller gain based on the phase margin – in digital control systems, the delay (T_d) component, including control, sampling, and PWM delays, is inevitable and compresses the phase margin of the open-loop transfer function. Thus, to preserve system stability, the derived upper limit of the controller gain needs to maintain a phase margin exceeding a set value, $PM_{\_min}$, which is given by

$$\begin{aligned} PM(\omega_{cross}) &= \pi + \angle G'_{u_i - i_g}(j\omega_{cross}, L_g) - \omega_{cross} T_d \\ &= \pi + \angle G'_{u_i - i_g}(j\omega_{cross}, L_g) - \frac{k_p |G_{in}(j\omega_{cross}) H(j\omega_{cross})|}{L_1 + L_2 + L_g} T_d \\ &\geq PM_{\_min} \end{aligned} \tag{6.10}$$

where ω_{cross} is the crossover frequency in hertz, and $\angle G'_{u_i - i_g}(j\omega, L_g)$ is the phase delay caused by $\angle G'_{u_i - i_g}(j\omega, L_g)$. Hence, from (6.10), the maximum loop gain of $K_{p\_max\_2}$ can be determined from the expected phase margin.

6.3 *E-Q*-Factor-Based Passive Damping Design

6.3.1 Principle of the Equivalent *Q*-Factor Method

Figure 6.3 shows the frequency responses of a high-order filter system without and with the passive damper. It is understood that the passive damper can reduce the *Q*-factor at the dominant resonant frequency, f_{res} (in hertz). For a high-order filter-based system, the optimized *Q*-factor is difficult to obtain by direct analysis of the

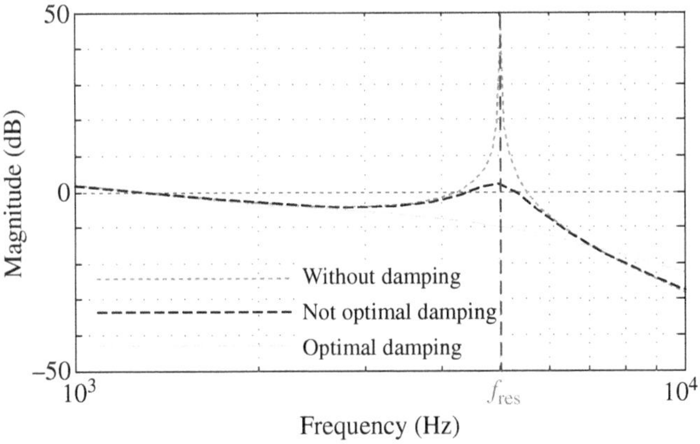

Figure 6.3 Bode plots of filters with different *Q*-factor values.

complex conjugate solutions of the transfer function $G'_{u_i-i_g}(s, L_g)$ because the dominant resonance frequency varies according to different parameters [6]. However, in [6], it was noted that the *LLCL* filter has almost the same frequency response as the *LCL* filter within half the switching frequency range.

Hence, similar to the *LCL* filter, the *LLCL* filter circuit can be simplified to an equivalent *LCR* series resonant circuit to calculate the equivalent *Q*-factor at the dominant resonance frequency. This method is referred to as *E-Q*-factor analysis and is given by

$$Q_E = \frac{1}{R_E}\sqrt{\frac{L_E}{C_E}} \tag{6.11}$$

where Q_E is the equivalent *Q*-factor, and R_E, L_E, and C_E are the equivalent resistance, inductance, and capacitance of the series *LCR* circuit, respectively. Three passively damped *LCL* and *LLCL* filter-based systems and their Bode diagrams for the grid-injected current versus output voltage of the inverter are shown in Figures 6.4, 6.5, and 6.6, respectively, and all their parameters are listed in

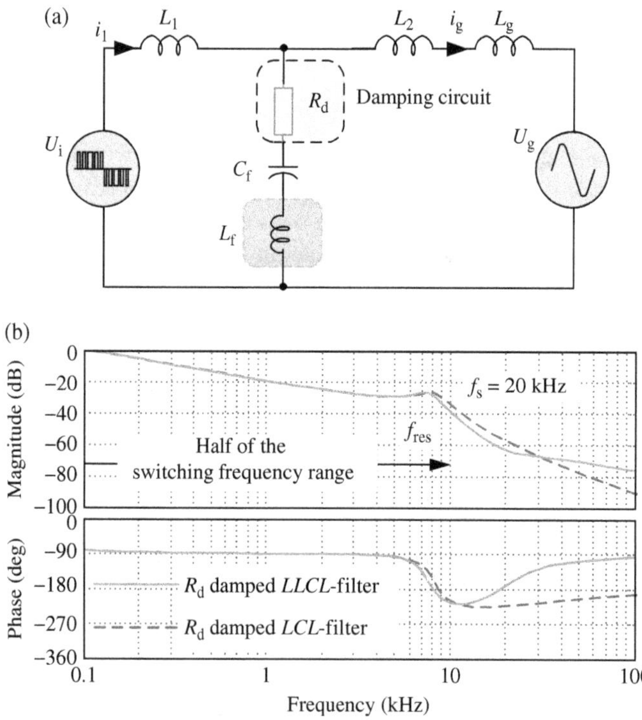

Figure 6.4 Simply damped *LCL* or *LLCL* filter with R_d-based inverter system: (a) topology and (b) Bode plots of transfer function $i_g(s)/u_i(s)$.

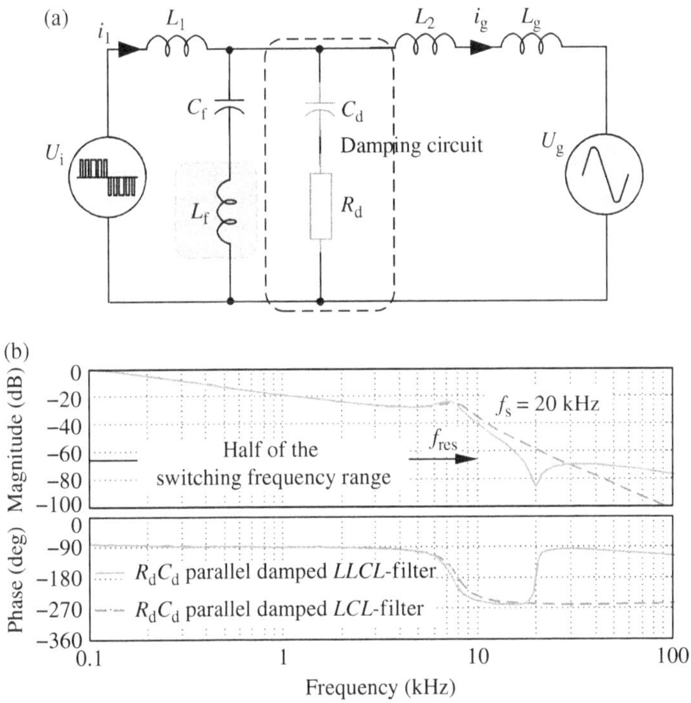

Figure 6.5 RC parallel damped *LCL* or *LLCL* filter-based inverter system: (a) topology and (b) Bode plots of transfer function $i_g(s)/u_i(s)$.

Table 6.1 with the grid inductance assumed as zero. It is further seen that the stability of the *LCL* or *LLCL* filter-based system is dependent on the dominant poles when using the *PR* + *HC* controller. Compared with the *LCL* filter, the *LLCL* filter does not involve problems with control of the entire inverter system.

Based on (6.11), the calculated *Q*-factors of the three passive damping methods are listed in Table 6.2 for $L_g = 0$.

6.3.2 *E-Q*-Factor-Based *RC* Parallel Damping Design

As shown in Figure 6.5a, if L_f is shortened, the RC parallel damped *LCL* filter can be obtained. The equivalent resistance R_E and capacitance C_E of the bypass capacitor C_f and RC parallel damper can be calculated as:

$$\left.\frac{1}{sC_f}//\left(\frac{1}{sC_d}=R_d\right)\right|_{s=j\omega} = \left.\frac{R_dC_ds+1}{(R_dC_dC_fs^2+(C_d+C_f)s)}\right|_{s=j\omega}$$

$$= R_E + \frac{1}{sC_E} \tag{6.12}$$

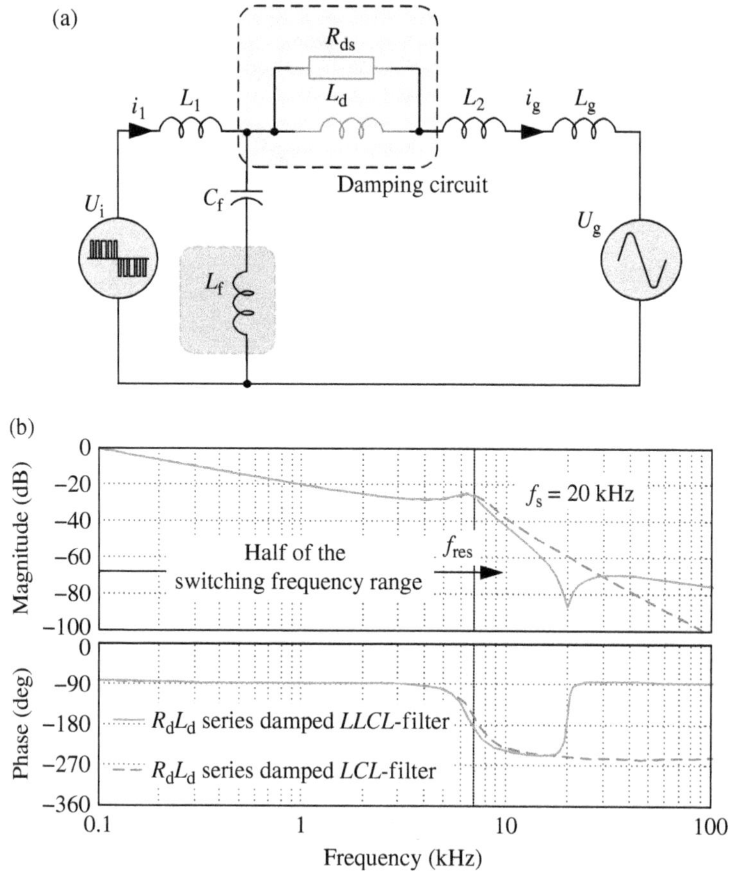

Figure 6.6 RL series damped *LCL* or *LLCL* filter-based inverter system: (a) topology and (b) Bode plots of transfer function $i_g(s)/u_i(s)$.

Then, at the dominant resonance frequency, the equivalent resistance R_E, inductance L_E, and capacitance C_E of the equivalent LCR circuit can be calculated as:

$$
\begin{aligned}
R_E &= \left.\frac{R_d C_d^2}{(C_d + C_f)^2 - (R_d C_d C_f s)^2}\right|_{s=j2\pi f_{res}} \\
L_E &= \frac{L_1(L_2 + L_g)}{L_1 + L_2 + L_g} \\
C_E &= \left.\frac{(R_d C_d C_f s)^2 - (C_d + C_f)^2}{R_d^2 C_d^2 C_f s^2 - (C_d + C_f)}\right|_{s=2\pi f_{res}}
\end{aligned}
\tag{6.13}
$$

Table 6.1 Parameters of the three passively damped filters.

Parameters of two filters					Parameters of dampers				
	LCL-filter	*LLCL*-filter	Parasitic resistance of the inductors			R_d damper	RC parallel damper	RL series damper	Composite damper
L_1	1.2 mH	1.2 mH	R_1	0.1 Ω	R_d	3 Ω	35 Ω	—	35 Ω
L_2	0.22 mH	0.22 mH	R_2	0.01 Ω	C_d	—	2 μF	—	2 μF
C_f	2 μF	2 μF	—	—	L_d	—	—	0.22 mH	0.22 mH
L_f	—	32 μH	R_f	0.2 Ω	R_d	—	—	7 Ω	7 Ω

Where the resistors R_1, R_2, and R_f are the respective parasitic resistance of the inductors L_1, L_2, and L_f.

Table 6.2 *Q*-factors of *LCL* and *LLCL* filters with three dampers.

	R_d damper	RC parallel damper	RL series damper
LCL-filter	3.214	3.978	3.603
LLCL-filter	3.479	3.742	4.102

Substituting (6.13) in (6.11), the *Q*-factor can be calculated at the dominant resonance frequency. To analyze the relationships between the *Q*-factor, damping resistor, and grid inductance, the normalization method is adopted. The base values of the impedance, inductance, capacitance, and resistance are defined as:

$$Z_b = \frac{(U_g)^2}{P_o} = R_b, R_b = \sqrt{\frac{L_b}{C_b}}, C_b = \frac{1}{\omega_o Z_b}, L_b = \frac{Z_b}{\omega_o} \tag{6.14}$$

where U_g is the rms value of the grid voltage, ω_o is the grid frequency in radians per second, and P_o is the active power generated by the inverter under the rated conditions.

Since R_d–C_d is in parallel with C_f, to balance the damping effect and damping losses, equal values of C_f and C_d are used [5]. If the discontinuous unipolar modulation mode is adopted and current ripple ratio of L_1 is 30%, the converter-side inductor can be derived as $L_1 = 0.0156L_b$. Assuming $L_2 = L_1$,

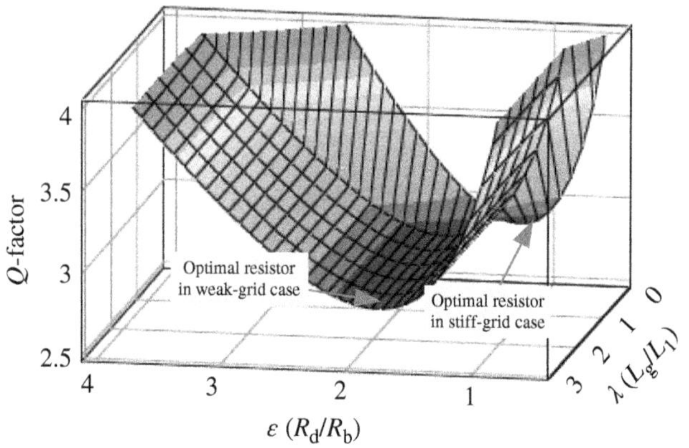

Figure 6.7 Q-factor as a function of ε and λ for the RC parallel damped LCL filter.

$L_g = \lambda L_1$, $C_f = C_d = 0.015C_b$, and $R_d = \varepsilon R_b$, the Q-factor can be plotted as a function of ε and λ, as shown in Figure 6.7. From Figure 6.7, it is seen that when the damping resistor is designed with optimal Q-factor for stiff grid applications, the Q-factor increases as the grid inductance increases; when the damping resistor is designed with optimal Q-factor for weak grid applications, the Q-factor will increase while the grid turns stiff. If the grid inductance varies over a wide range, this damping method cannot always achieve the optimal Q-factor.

From Figure 6.5, it is also seen that within half the switching frequency range, the RC parallel damped LCL and $LLCL$ filters have almost similar frequency response characteristics when the parameters are the same except for L_f. Hence, for the $LLCL$ filter-based inverter system, the E-Q-factor analysis method can be used to select a reasonable damping resistance. It should be noted that the RC parallel damped $LLCL$ filter cannot be used for large variations of the grid inductance [6].

6.3.3 *E-Q*-Factor-Based *RL* Series Damping Design

In [7], another passive damping method called the RL series damping method was introduced for the LC filter. In theory, this damping method is also effective for the LCL filter, as shown in Figure 6.6a, where L_f is shortened. Herein, the exact damping parameters and application of this damping method are discussed using the *E-Q-factor* analysis approach. By simplifying the complex high-order circuit topology to an equivalent series LCR circuit, the equivalent R_E and L_E of L_1, L_2, L_g, and the RL series damper can be calculated as:

$$\left.\left(\frac{sL_1\left(\dfrac{SL_dR_{ds}}{sL_d + R_{ds}} + sL_2'\right)}{sL_1 + \dfrac{SL_dR_{ds}}{sL_d + R_{ds}} + sL_2'}\right)\right|_{s=j\omega} = R_E + sL_E \tag{6.15}$$

where $L_2' = L_2 + L_g$. Then, the equivalent resistance R_E, inductance L_E, and capacitance C_E at the dominant resonant frequency in the series LCR circuit are obtained as:

$$R_E = \left.\frac{-s^2L_1^2L_d^2R_{ds}}{\left(L_1 + L_2' + L_d\right)^2R_{ds}^2 - S^2\left(L_1 + L_2'\right)^2L_d^2}\right|_{s=j2\pi f_r}$$

$$L_E = \left.\frac{L_1\left(L_2' + L_d\right)\left(L_1 + L_2' + L_d\right)R_{ds}^2 - s^2L_1L_2'\ \left(L_1 + L_2'\right)L_d^2}{\left(L_1 + L_2' + L_d\right)^2R_{ds}^2 - s^2\left(L_1 + L_2'\right)^2L_d^2}\right|_{s=j2\pi f_r}$$

$$C_E = C_f \tag{6.16}$$

Assuming $\delta = L_d/L_2'$, the relationship between the Q-factor, δ, and η ($\eta = R_{ds}/R_b$) is plotted in Figure 6.8 under the conditions that $L_1 = L_2 = 0.0156L_b$, $C_f = 0.015C_b$, and $L_g = 0$, which shows that for larger δ, the Q-factor is better. However, a large damping inductor increases the damping loss and system cost. Since L_g is in series with L_2, a larger L_g will also result in worsened Q-factor.

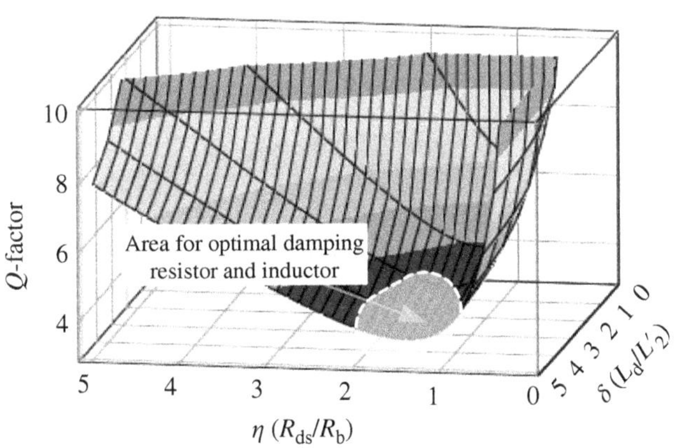

Figure 6.8 Q-factor as a function of δ and η for the RL series damped *LCL* filter.

For the *LLCL* filter, the RL series damping method is also effective in theory. Compared with the *LCL* filter, the difference here is that the equivalent inductor of the *LLCL* filter needs to be $(L_E + L_f)$. Owing to the smaller grid-side inductor, both the damping inductance and damping losses are smaller than those of the RL series damped *LCL* filter. It is noted that the RL series damping method is only effective under stiff grid conditions for both the *LCL* and *LLCL* filters.

6.4 New Composite Passive Damping Scheme for the *LLCL* Filter

6.4.1 Composite Passive Damping Scheme

As noted previously, for the *LLCL* filter-based system, the RC parallel damping method is suitable for grids with relatively narrow ranges of inductance values, whereas the RL series damping is suited only for stiff grid conditions. To ensure stability of the grid-connected inverter system for grids with wide variations of the inductance, a composite damping method for the *LLCL* filter is proposed, as shown in Figure 6.9, at the expense of additional power loss and total inductance.

The design requirements of the damping parameters are summarized as follows:

1) Damping resistor R_d needs to be designed for optimal *Q*-factor under the weakest grid condition while C_d is equal to C_f;
2) damping resistor R_{ds} is selected for optimal *Q*-factor under the stiffest grid condition while L_d is equal to L_2.

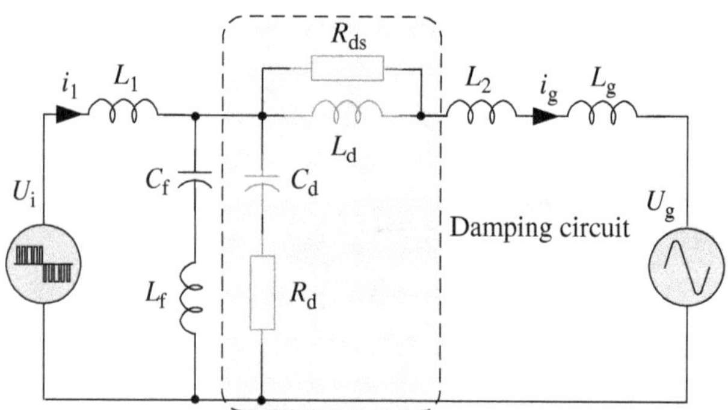

Figure 6.9 Schematic of the composite damping method.

6.4.2 Design Example

When $f_s = 20$ kHz, $U_{dc} = 350$ V, $U_g = 220$ V/50 Hz, $P_{rated} = 2$ kW, and the discontinuous unipolar modulation mode, the main parameters of the *LLCL* filter are designed based on the criteria in [2], which are also listed in Table 6.1:

1) Only with RC parallel damper: The *Q*-factor of the RC parallel damped *LLCL* filter as a function of the damping resistor and grid inductance is plotted in Figure 6.10 for Cd = C_f.

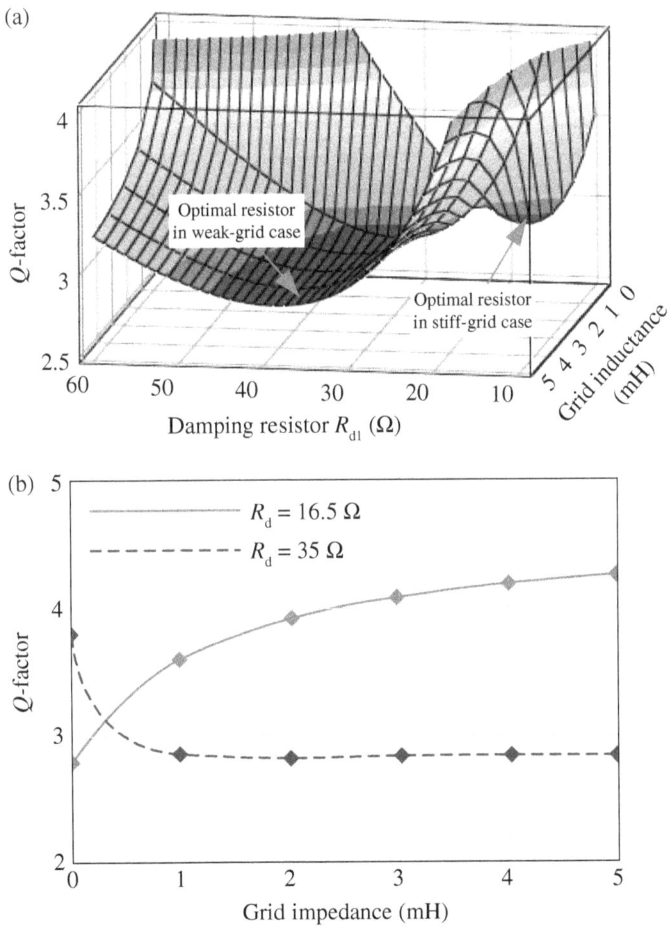

Figure 6.10 *Q*-factor of the RC parallel damped *LLCL* filter. (a) Relationships between *Q*-factor, damping resistor, and grid inductance. (b) *Q*-factor versus grid inductance for R_d = 16.5 and 35 Ω.

From Figure 6.10a, it is seen that relatively increasing damping resistances are needed to achieve optimal Q-factor with increasing grid inductances. To show the influence of inductance variations on Q-factor, Figure 6.10b depicts that the damping resistance of 16.5 Ω, which is designed for optimal Q-factor under stiff grid conditions, has less damping effect with increasing grid inductance. On the contrary, when the damping resistor is selected as 35 Ω, which is suitable for the situation with L_g=5 mH, the Q-factor exceeds 4 if L_g is close to 0 mH. It is difficult to select the damping resistor of the RC parallel damper with optimal Q-factor if the grid inductance varies over a wide range.

2) Only with RL series damper: If $L_d = L_2$, the Q-factor as a function of the damping resistor and grid inductance is plotted in Figure 6.11. It is seen that the larger grid inductance, the greater is the Q-factor and worse are the damping effects achieved. Hence, this damping method is only effective in the stiff grid. The damping resistor can be calculated by analyzing the first-order derivative of the equivalent Q-factor; a damping resistance of 7 Ω seems to achieve good damping effects for the designed $LLCL$ filter under the stiffest grid condition ($L_g = 0.15$ mH).

3) Composite passive damping scheme: For the composite damped $LLCL$ filter, when the damping parameters are selected according to the design method introduced earlier, it is seen that a Q-factor of less than 3 can be achieved over a wide range of grid inductance values, as shown in Figure 6.12. It is noted that the stability does not depend only on the Q-factor, especially when considering the loop delay. However, a small Q-factor always helps the stability of the entire system.

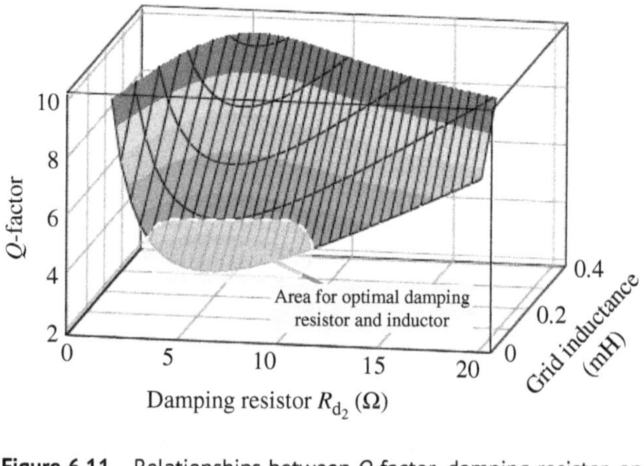

Figure 6.11 Relationships between Q-factor, damping resistor, and grid inductance for the RL series damped $LLCL$ filter.

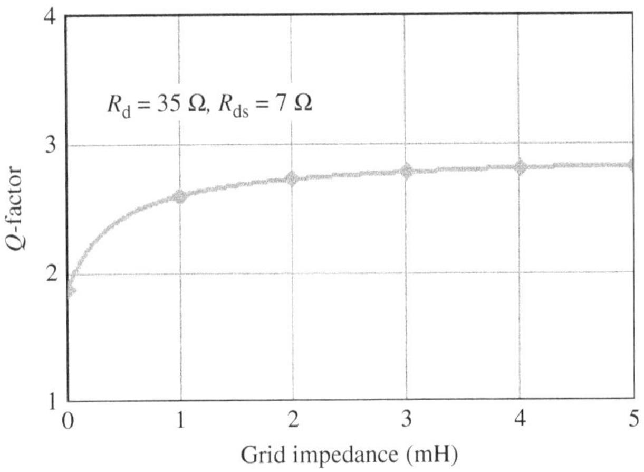

Figure 6.12 *Q-factor* of the composite damped *LLCL* filter.

6.4.3 Analysis of Achieved Damping

The damping parameters of the three passively damped *LLCL* filters are listed in Table 6.1, where the current sensor gain and gain of the PWM inverter are 0.0182 and 1400, respectively (same as those of the experimental prototype). To insert the ninth harmonic compensator in the weakest grid condition ($L_g = 5$ mH), the minimum controller gain should exceed 0.73, according to (6.4). From (6.7) and (6.8), $k_{p\_max\_1}$ is calculated as 0.76, and the corresponding grid inductance is about 0.65 mH. In a real system with the digital controller, control delays are inevitable. However, with appropriate digital signal processing (DSP) control, the total delay of the system can be reduced to 0.75 T_s [8] or less [9] (T_s is the switching period of 50 μs). Herein, the control delay is selected as 0.75 T_s. Using (6.10), when the phase margin is set to 45°, $k_{p\_max\_2}$ can be calculated as 0.81 under the stiffest grid condition ($L_g = 0.15$ mH). Then, according to (6.5), the final proportional gain k_p of the *PR* + *HC* controller is selected as 0.76, and the control bandwidth of 2.5 kHz under the stiffest and 520 Hz under the weakest grid conditions can be achieved theoretically.

The *PR* + *HC* controller is expressed as (6.17), where k_p is the proportional gain and K_{ih} represents the individual resonant integral gain:

$$G_{PR}(s) = k_p + \sum_{h=1,3,5,7,9} \frac{k_{ih}s}{s^2 + (\omega_o h)^2} \tag{6.17}$$

Here, K_{ih} is selected as 100 for each harmonic that is compensated. The Bode plots of the three passively damped filters with the parameters given in Table 6.2 and designed controller are shown in Figure 6.13. It is seen that the

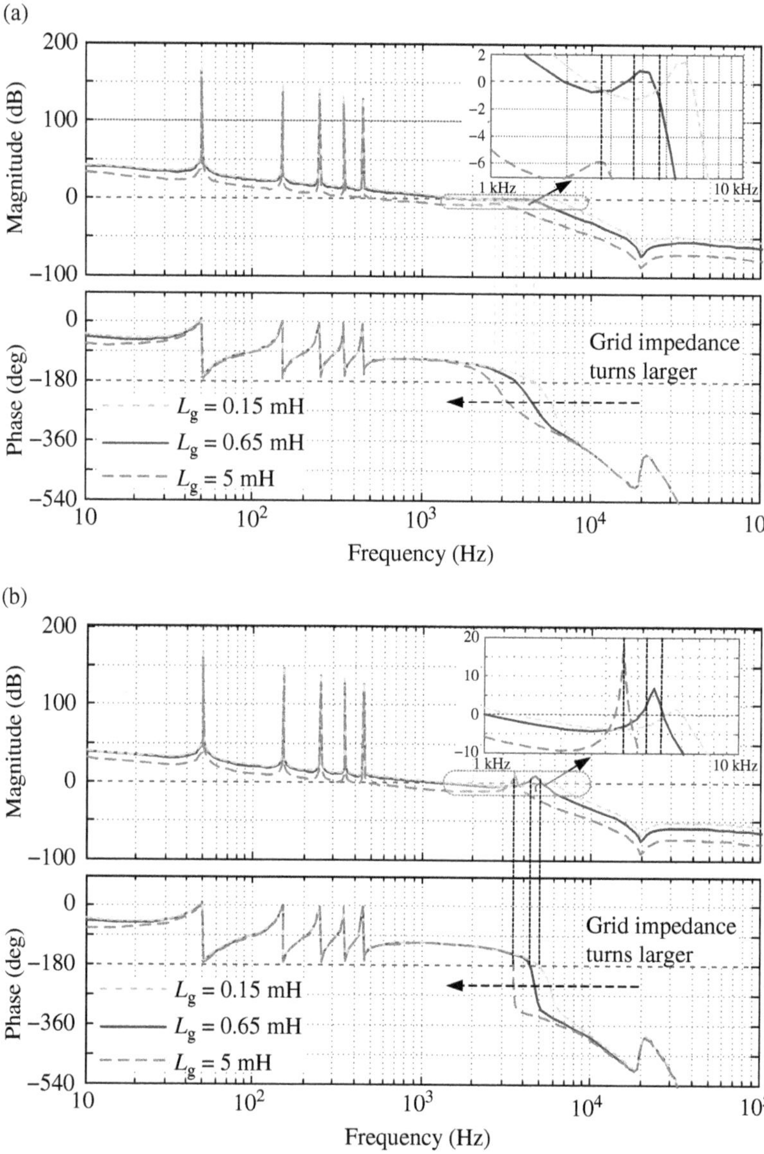

Figure 6.13 Bode plots of $G_{open-loop}(j\omega)$ with a delay of 0.75 T_s for (a) Case 1: only the RC parallel damper; (b) Case 2: only the RL series damper; and (c) Case 3: with the composite damper.

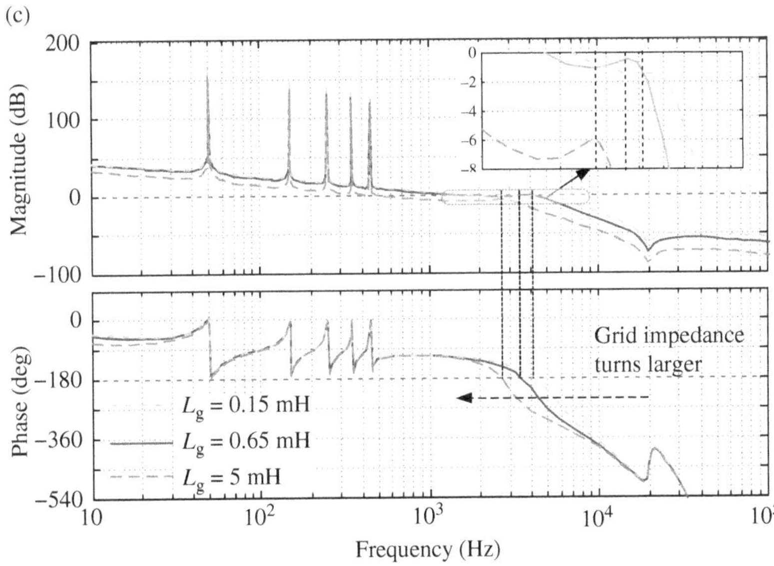

Figure 6.13 (Continued)

system with only the RC parallel damper (Figure 6.13a) is unstable when the grid inductance is 0.65 mH, the system with the RL series damper (Figure 6.13b) is stable only under the stiff grid condition, and the system with the composite damper (Figure 6.13c) can maintain stability over a wide range of grid inductances. Therefore, in terms of acceptable system stability, the composite damping method may be the best out of the three damping methods.

To confirm the effectiveness of the proposed damping method, simulations were performed using the PSIM (Power Simulation) software. The parameters used are the same as the designed values. Referring to the design procedure given in Sections 6.3 and 6.4.3, simulations are conducted for the three cases under the conditions $f_s = 20$ kHz, $U_{dc} = 350$ V, $U_g = 220$ V/50 Hz, $P_o = 2$ kW, and $T_d = 0.75\ T_s$. The current sensor gain and gain of the PWM inverter are 0.0182 and 1400, respectively, for all cases; a *PR* + *HC* (from third to ninth) controller is selected for grid-current feedback. The parameters of the *PR* + *HC* controller are the same as those used in the analyses.

Figure 6.14 shows the system diagram for the simulations and experiments, where the PCC represents the common connection point. The circuit in the dashed lines is used to emulate the grid with varying inductance values. At time t_0, S_1 switches ON, and the inverter works in the offline state. At time t_1, S_2 is ON, and the inverter changes from the offline to online state. For simulation, it is

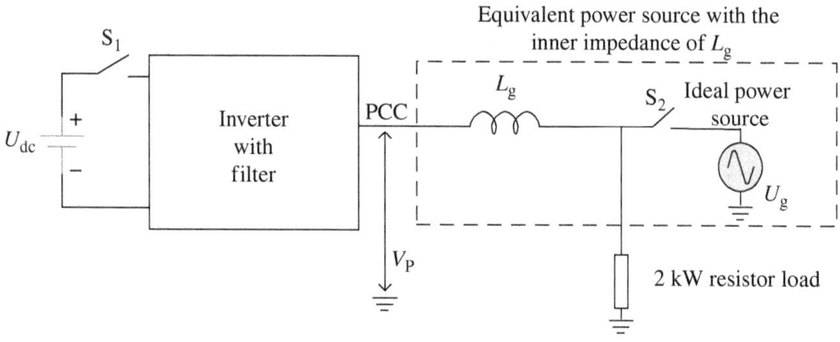

Figure 6.14 System diagram for the simulations and experiments.

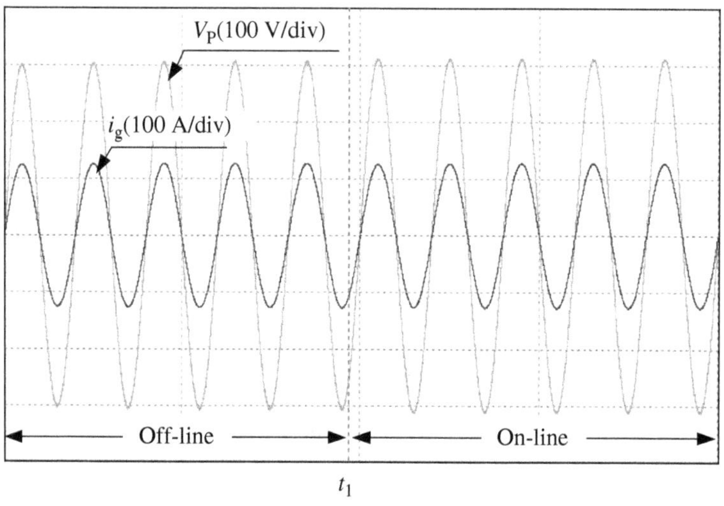

Figure 6.15 Simulated grid-injected currents and PCC voltages in the stable state.

assumed that the system operates from the offline state directly. When the system is stable, the simulated waveforms of the grid-injected current and PCC voltage are similar, as shown in Figure 6.15.

Figure 6.16 shows that when $L_g = 5$ mH, the *LLCL* filter-based system with only the RL series damper is not stable in both the offline and online states. Figure 6.17 shows that when $L_g = 0.65$ mH, the *LLCL* filter-based system, regardless of the RC parallel or RL series dampers, cannot maintain stability in the online state. These simulations match the theoretical analyses well.

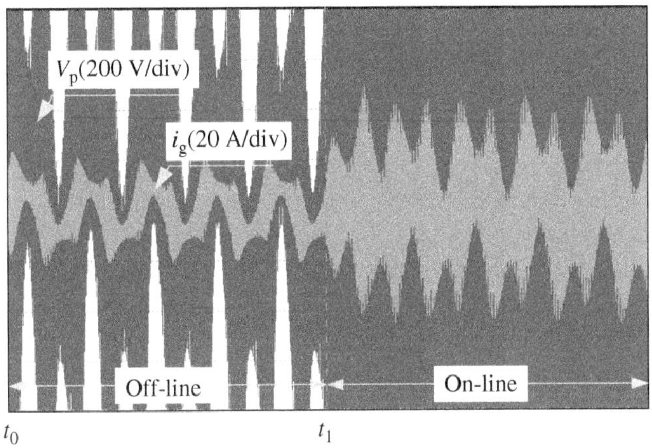

Figure 6.16 Simulated grid-injected currents and PCC voltages under conditions of delay of 0.75 T_s, L_g = 5 mH, and using only the RL series damper.

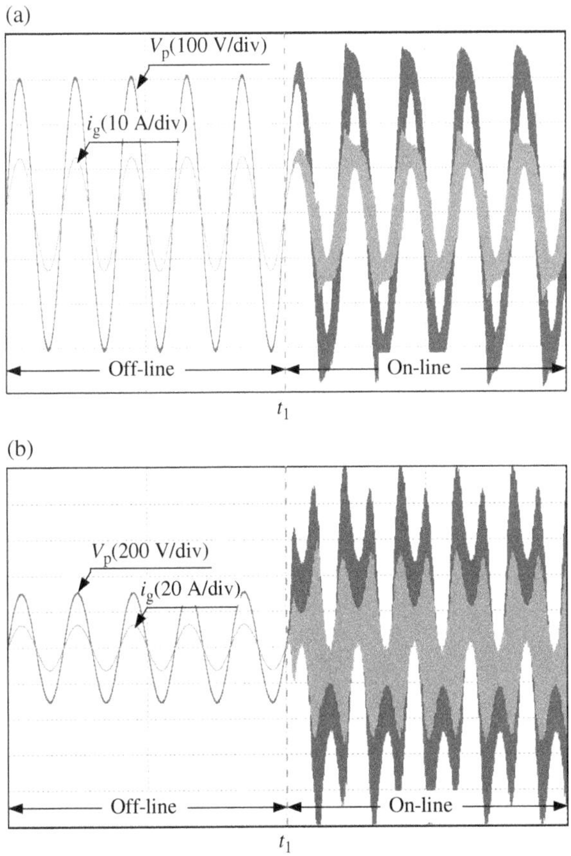

Figure 6.17 Simulated grid-injected currents and PCC voltages under the condition of delay of 0.75 T_s, L_g = 0.65 mH with only the (a) RC parallel and (b) RL series dampers.

6.5 Experimental Verification

To verify the theoretical observations, a 2 kW prototype device based on a DSP (TMS320LF2812) controller was constructed, and its system diagram is shown in Figure 6.14, where a programmable ac source (Chroma 6530) is used to emulate ideal grid voltages. The parameters of the filters are the same as those noted previously, and the three different damping methods are evaluated under the conditions $f_s = 20$ kHz, $U_{dc} = 350$ V, $U_g = 220$ V/50 Hz, $P_{rated} = 2$ kW, dead time of 2 μs, and delay of 0.75 T_s. The measured PCC voltage waveforms and grid-injected currents are similar in the stable state, as shown in Figure 6.18.

Figure 6.19 shows the experimental results under the weak grid condition of $L_g = 5$ mH and delay of 0.75 T_s, showing that the RL series damped system cannot maintain stability in both the offline and online states. It is also seen that owing to resonance, the overcurrents trigger the hardware protections at time t_2. Note that owing to hardware protection, the state transition (from offline to online at time t_1) of the RL series damped system cannot be observed directly. As shown in Figure 6.19a, b, the resonances in the online and offline states are observed using switch S_1, which is ON at time t_0.

Figure 6.20 shows that when $L_g = 0.65$ mH, the *LLCL* filter-based system with only the RC parallel or RL series damper cannot maintain stability in the online state.

Figure 6.21a shows the measured grid-injected currents and grid voltages of the composite damping *LLCL* filter-based grid-connected inverter under the condition of $L_g = 5$ mH, with the magnitudes of the injected 3rd, 5th, 7th, 9th, and 11th

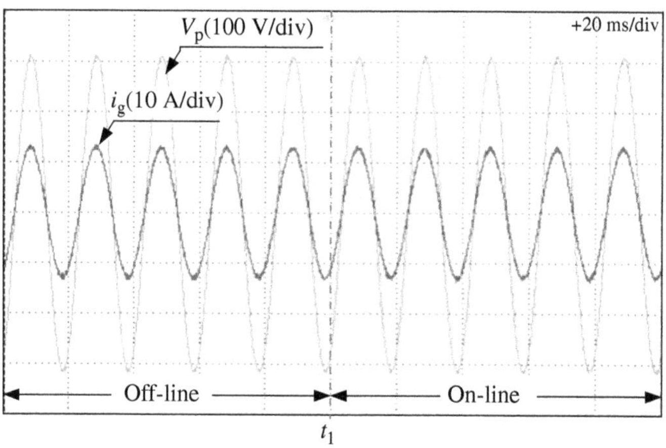

Figure 6.18 Measured grid-injected currents and PCC voltages in the stable state.

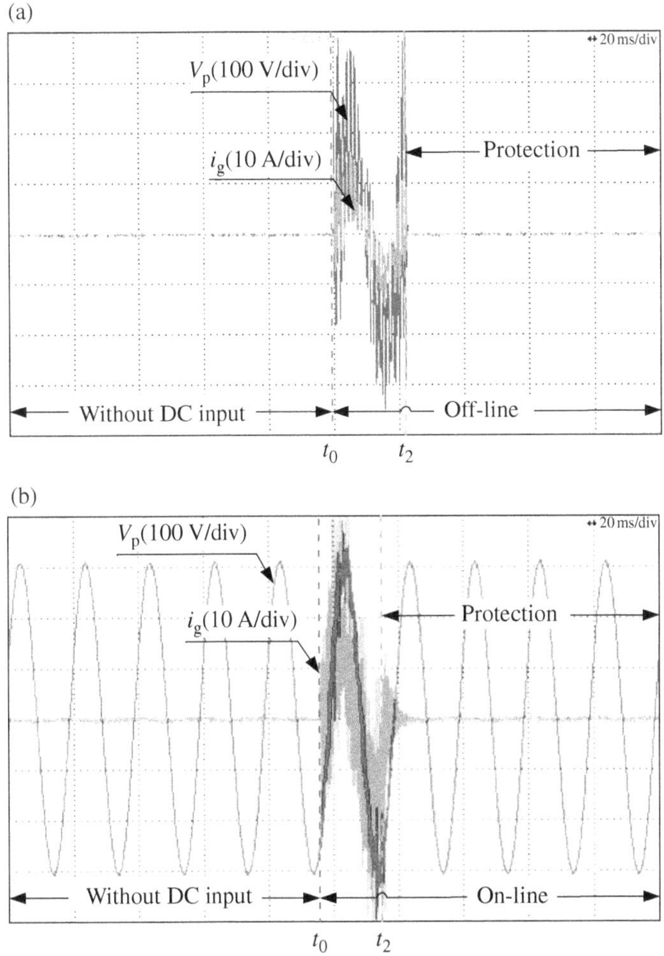

Figure 6.19 Measured grid-injected currents and PCC voltages under the condition of delay of 0.75 T_s, L_g = 5 mH, with only the RL series damper in the (a) offline and (b) online states.

harmonics with respect to the fundamental component of U_g being 1.2%, 2.8%, 1.3%, 2.4%, and 1.5%, respectively. Figure 6.21b shows the spectra of the grid-injected current and grid voltage. Since harmonic compensations of the 3rd, 5th, 7th, and 9th components are adopted, the related harmonic currents are suppressed well, while the 11th harmonic current cannot be alleviated. The experimental results show that under given grid conditions with large inductance variations, the control robustness of the system with the composite damper is better compared with that of the systems with only the RC parallel or RL series

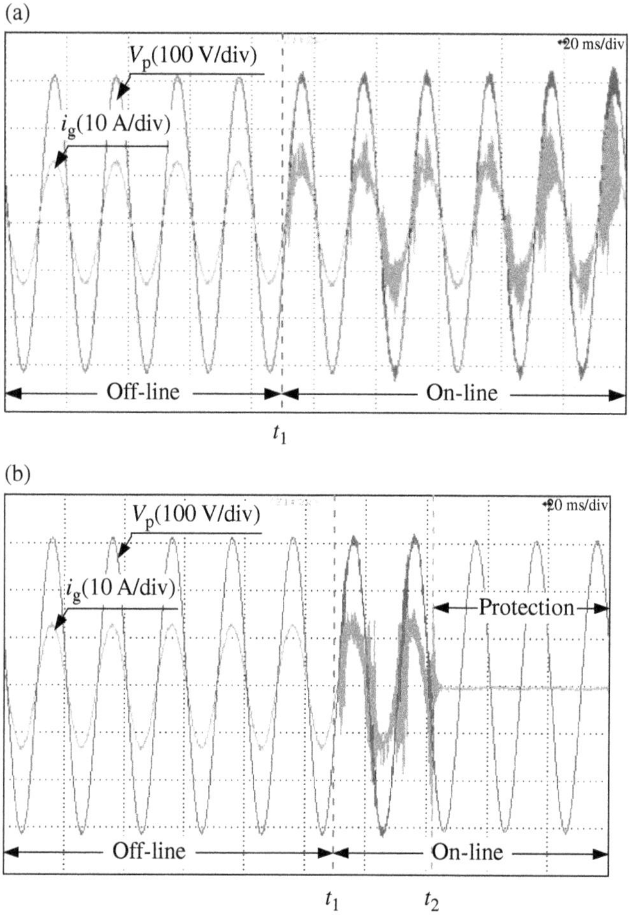

Figure 6.20 Measured grid-injected currents and PCC voltages under the condition of delay of 0.75 T_s, L_g = 0.65 mH, with only the (a) RC parallel and (b) RL series dampers.

damper. These experimental results match the theoretical analyses and simulations well.

6.6 Summary

This chapter discusses the passive damping design for high-order power filter-based grid-connected inverters. Based on the simulations and experiments, the following conclusions can be drawn:

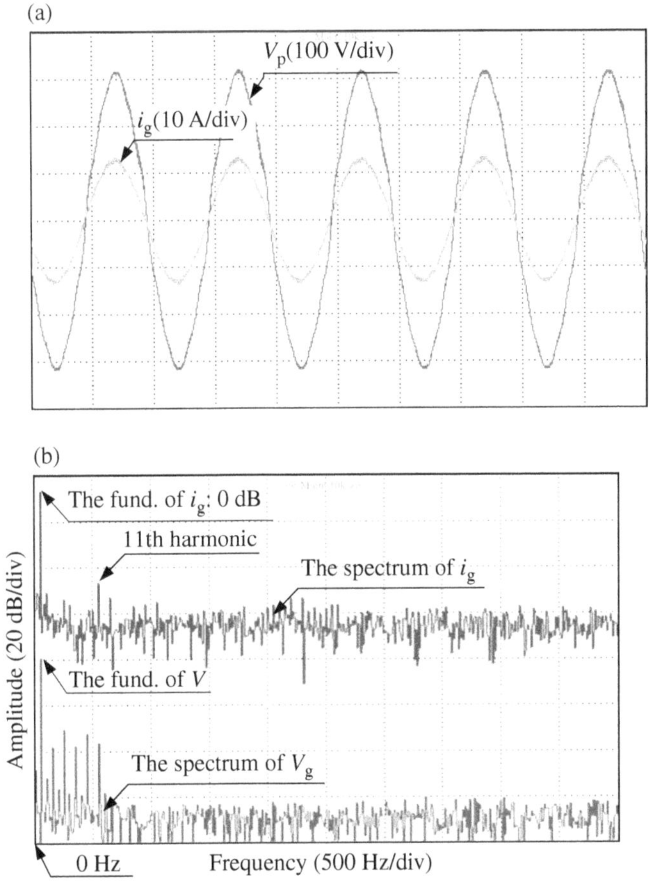

Figure 6.21 Measured grid-injected current and grid voltage under the condition of delay of 0.75 T_s, L_g = 5 mH, and composite damper: (a) grid-injected current and distorted grid voltage; (b) spectra of the grid-injected current and grid voltage.

1) The RC parallel damping method can be adopted for systems under weak or stiff grid conditions but is only suitable for grids with narrow inductance variations;
2) the RL series passive damping method is useful only for systems under the stiff grid condition;
3) a robust passive damping method can be achieved for the *LLCL* filter-based inverter connected to a grid with large inductance variations with a composite passive damper, where its RC parallel part is designed under the weakest grid condition and RL series part is designed under the stiffest grid condition, thus incurring additional cost of materials and power losses;

4) the proposed damping method may only be suitable for *LLCL* filter-based systems owing to the additional power losses and inductance.

The effectiveness of the proposed damping method has been fully verified via simulations and experiments with a 2 kW *LLCL* filter-based single-phase grid-connected inverter prototype with fixed controller gain, while varying the grid inductance from 0.15 to 5 mH.

References

1 Lindgren, M. and Svensson, J. (1995). Connecting fast switching voltage source converters to the grid-harmonic distortion and its reduction. *Proceedings of the IEEE Stockholm Power Technology Conference*, pp. 191–196, Stockholm, Sweden (18–22 June). New York: IEEE.

2 Wu, W., He, Y., and Blaabjerg, F. (2012). An *LLCL*- power filter for single-phase grid-tied inverter. *IEEE Trans. Power Electron.* 27 (2): 782–789.

3 Yuan, X., Merk, W., Stemmler, H., and Allmeling, J. (2002). Stationary-frame generalized integrators for current control of active power filters with zero steady-state error for current harmonics of concern under unbalanced and distorted operating conditions. *IEEE Trans. Ind. Appl.* 38 (2): 523–532.

4 Liserre, M., Teodorescu, R., and Blaabjerg, F. (2006). Stability of photovoltaic and wind turbine grid-connected inverters for a large set of grid impedance values. *IEEE Trans. Power Electron.* 21 (1): 263–272.

5 Channegowda, P. and John, V. (2010). Filter optimization for grid interactive voltage source inverters. *IEEE Trans. Ind. Electron.* 57 (12): 4106–4114.

6 Wu, W., He, Y., and Blaabjerg, F. (2013). A new design method for the passive damped LCL- and LLCL-filter based single-phase grid-tied inverter. *IEEE Trans. Ind. Electron.* 60 (10): 4339–4350.

7 Xing, L., Feng, F., and Sun, J. (2011). Optimal damping of EMI filter input impedance. *IEEE Trans. Ind. Appl.* 47 (3): 1432–1440.

8 Grogan, S.A.S., Holmes, D.G., and McGrath, B.P. (2011). High-performance voltage regulation of current source inverters. *IEEE Trans. Power Electron.* 26 (9): 2439–2448.

9 Corradini, L., Stefanutti, W., and Mattavelli, P. (2008). Analysis of multisampled current control for active filters. *IEEE Trans. Ind. Appl.* 44 (6): 1785–1794.

Part IV

Robust Control Design

7

Robust Hybrid Damper Design for *LCL/LLCL*-Filtered Grid-Connected Inverter

7.1 Introduction

In "Composite passive damping scheme for *LLCL*-filtered grid-connected inverter," we discussed the bases of the upper and lower limits of the $PR + HC$ controller gain and the equivalent Q-factor (E-Q-factor) calculation method; moreover, a robust passive damping scheme was proposed for the *LLCL* filter-based grid-connected inverter to overcome the adverse effects of large grid-inductance variations and to suppress resonance [1]. Background harmonic voltages (BHVs) may distort the injected currents of grid-connected inverters. To resist the effects of BHVs and ensure that the grid currents meet the required standards, the effects of the grid voltages on grid currents should be mitigated. Thus, extra dampers are inserted to maintain system stability, which can be achieved in three ways. The first approach is the grid-voltage feedforward method, which is widely used to suppress low-order harmonics of grid-injected currents [2–4]. The second approach is to insert a proportional resonant regulator with a harmonic compensator ($PR + HC$ regulator) in the grid-connected inverter using a low control bandwidth [5], where the lowest frequency of the HC exceeds the control bandwidth under any given grid condition. The third approach involves adopting the $PR + HC$ regulator for a grid-connected inverter based on a high-bandwidth control strategy [6–8].

Inverters with capacitor-current-feedback active damping (AD) are presented in this chapter because of their efficiency and flexibility. There are two main types of ADs. The first is the single-loop filter-based AD, where a notch filter is used to attenuate the resonance of the high-order filter [9, 10]. Owing to variable equivalent grid impedances, a self-commissioning notch filter was proposed in [10]; however, this method relies on the accuracy of real-time resonance frequency estimation, which is inevitably affected by the harmonics at the point of common

Control and Filter Design of Single-Phase Grid-Connected Converters, First Edition. Weimin Wu, Frede Blaabjerg, Henry Chung, Yuanbin He, and Min Huang.

Published 2023 by John Wiley & Sons, Inc.

coupling (PCC). The other type of AD involves multiloop control, such as a virtual resistor [5, 11], capacitor-voltage feedback [12], capacitor-current feedback [5, 13–17], and multivariable feedback [18]. For a multiloop AD method, the computational delay of the inner loop may result in system poles on the right half of the complex plane, which complicates the controller design and system stability [11].

The layout of this chapter is as follows. First, the bandwidth of the grid-current feedback control is analyzed by considering a digital delay and the single-loop hybrid damping method for *LCL* or *LLCL* filter-based single-phase grid-connected inverters, followed by the stepwise design and example illustration of the proposed method. Experimental results are then presented for a 2 kW single-phase grid-connected inverter prototype to confirm the effectiveness of the proposed method. Finally, some conclusions of the study are presented.

7.2 Control Bandwidth Analysis of the Grid-Current Feedback Method

7.2.1 *LCL/LLCL*-Filtered Grid-Connected Inverter System

The capacitor-current feedback is one of the classical damping methods for the high-order filter-based grid-connected inverter. Figure 7.1 shows the configuration of a single-phase *LCL* or *LLCL* filter-based grid-connected voltage source inverter with the capacitor-current-feedback AD method, where we have the following.

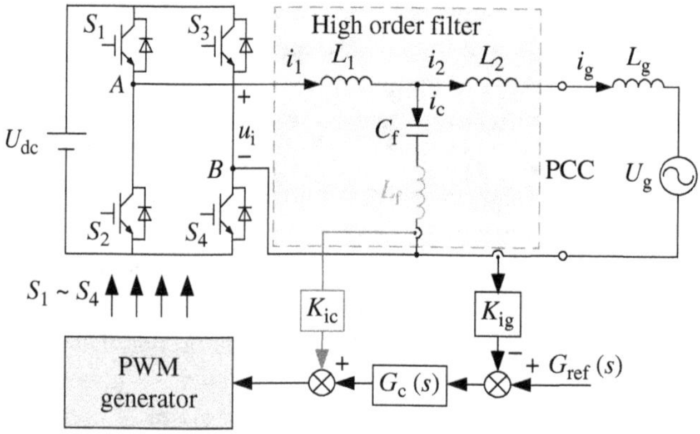

Figure 7.1 Topology of a single-phase high-order-filter-based grid-connected voltage source inverter with the capacitor-current-feedback active damping method.

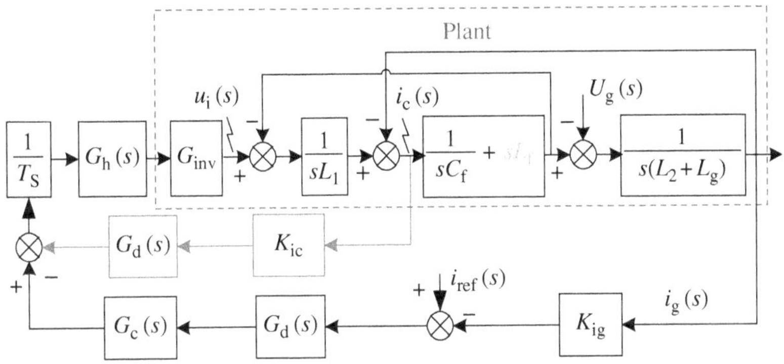

Figure 7.2 Topology of a single-phase high-order-filter-based grid-connected voltage source inverter with the capacitor-current-feedback AD method.

- Inverter-side inductor L_1, filter capacitor C_f, and grid-side inductor L_2 give the *LCL* filter. If series inductor L_f in the bypass capacitor loop is inserted, the *LCL* filter will be changed into the *LLCL* type.
- L_g is the equivalent grid impedance.
- K_{ic} and K_{ig} are the sensor gains of the capacitor and the grid current, respectively.
- $G_c(s)$ is the designed controller, which is the *PR* + *HC* regulator in this paper.

The block diagram of the digitally controlled *LLCL* filter-based grid-connected inverter is depicted in Figure 7.2, where G_{inv} is the transfer function of the PWM inverter, and it is the ratio of the input dc voltage U_{dc} and the amplitude of the carrier U_{tri} [13]. When L_f =0, the *LLCL* filter can be transformed into the *LCL* filter.

7.2.2 Maximum Achieved Bandwidth of the Control Method

The system open-loop gain $G_{open\text{-}loop\_c(s)}$ can be described as (7.1), shown as following:

$$
\begin{aligned}
G_{open\text{-}loop\_c}(s) &= \frac{i_g(s)}{i_{ref}(s)} \\
&= \frac{K_{ig}G_{inv}G_c(s)(1+s^2L_fC_f)e^{-sT_d}}{s^3\left[L_1(L_2+L_g)+L_1L_f+L_f(L_2+L_g)\right]C_f + s^2\left(K_{ic}G_{inv}(L_2+L_g)C_f\right)e^{-sT_d}+s(L_1+L_2+L_g)}
\end{aligned} \tag{7.1}
$$

According to (7.1), if $L_f = 0$, the phase margin of the *LCL* filter-based grid-connected inverter system can be calculated with (7.2). The delay can be transformed into the following using Euler's formula:

$$\mathrm{PM_c} = 180^\circ + \angle \frac{K_{ig}G_{inv}G_c(s)e^{-sT_d}}{s^3L_1L_2C + s^2L_2CK_{ic}G_{inv}e^{-sT_d} + s(L_1 + L_2)}\bigg|_{s=j2\pi f_c} \tag{7.2}$$

$$e^{-T_ds} = \cos\left(\varphi_{delay}\right) - j\sin\left(\varphi_{delay}\right) \tag{7.3}$$

and the phase lag caused by overall delay T_d (mainly composed of computational and PWM delays) can be expressed as:

$$\varphi_{delay} = T_d \cdot 360 \cdot f \tag{7.4}$$

When $\varphi_{delay} < 90°$, substituting (7.3) in (7.2) gives the following phase margin:

$$\mathrm{PM} = \arctan \frac{2\pi L_1\left(f_r^2 - f_c^2\right) - K_{ic}G_{inv}f_c\sin\left(\varphi_{delay}\right)}{K_{ic}G_{inv}f_c\cos\left(\varphi_{delay}\right)} - \varphi_{delay}\bigg|_{f=f_c} \tag{7.5}$$

where resonance frequency $f_r = (1/2)\sqrt{(L_1 + L_2 + L_g)/(L_1(L_2 + L_g)C_f)}$ and f_c is the crossover frequency. Obviously, we have,

$$\mathrm{PM} \le 90 - \varphi_{delay} \tag{7.6}$$

When $\varphi_{delay} = 90°$, PM = 0, which means that the phase of the loop gain will be $-180°$ when $f = 1/(4T_d)$.

To ensure a 45°PM, the phase lag caused by the control delay should be limited to 45°. It can be calculated with (7.4) and (7.5) that the maximum crossover frequency is $f_{cmax} = 1/(8T_d)$ Hz.

As discussed in [16], the effect of the filter capacitor is negligible when calculating the magnitude of the loop gain in the frequency range of $[0, f_c]$. Meanwhile, the *PR* compensator can be simplified to a proportional element K_P at f_c. Thus, the magnitude of $G_{open\text{-}loop\_c(s)}$ at f_c can be approximated to:

$$\left|G_{open\text{-}loop\_c}(s)\right|\Big|_{s=j2\pi f_c} = \left|\frac{G_{inv}K_PK_{ig}}{s(L_1 + L_2 + L_g)}\right|\Bigg|_{s=j2\pi f_c} \tag{7.7}$$

It can be found from (7.7) that the bandwidth is inversely proportional to the total inductance. Note that the equivalent grid impedance varies within a wide range in the application due to the different numbers of grid-connected inverters in parallel at the PCC. In this paper, a weak grid coefficient η is defined to directly show the influence of the grid impedance variation as follows:

$$\eta = \frac{L_1 + L_2 + L_{g\,max}}{L_1 + L_2 + L_{g\,min}} \tag{7.8}$$

Then, together with the maximum crossover frequency of f_{cmax} under the stiffest grid condition, the maximum crossover frequency f_{cmax_w} under the weakest grid condition can be derived as follows:

$$f_{c\max\text{-w}} = \frac{f_{c\max}}{\eta} \tag{7.9}$$

An example is analyzed here to obtain a reasonable η. There is a 40 kW system, and 20 sets of 2 kW/220 V/50 Hz inverters are placed in parallel and connected to the grid power transformer (40 kVA) through a 200 m power line. The grid impedance mainly consists of short-circuit impedance Z_g and cable impedance Z_l, which can be chosen as follows:

1) For a 40 kVA transformer, short-circuit impedance Z_g is 0.19 mH (5%).

In the absence of other information, the cable reactance is selected as 0.08 mΩ/m for 50 Hz systems. The cable reactance is considered a pure inductive reactance here, which is 0.25 μH/m. A 200 m line means that Z_l is 0.05 mH.

The minimum value L_{gmin} is supposed to be 0 mH when the inverter is in the island mode. When all the paralleled inverters are under operation, the largest equivalent grid impedance L_{gmax} will be 4.8 mH. Therefore, the equivalent grid impedance varies from 0 to 4.8 mH. When $L_1 = 1.2$ mH, $L_2 = 0.22$ mH, $L_f = 2$ μF, and the sampling frequency of the controller is 20 kHz, the maximum value of η is calculated as 4.4. In theory, if the total delay is 0.75 switching period, f_{cmax} and f_{cmax_w} will be 3334 and 756 Hz, respectively.

When the capacitor-current feedback coefficient $K_{ic} = 0$, the dual-loop current control changes to the single-loop current control. The effective current control bandwidth mainly depends on the total delay. It can be deduced that, as long as the total delay is the same, a similar bandwidth can be achieved for both single-loop and multiloop current controls. With the asymmetry sampling method, the delay time can be reduced to 0.75 switching period [19, 20]. However, due to the possible aliasing error and the switching noise caused by the capacitor current sampling, it is difficult for the dual-loop current control method to reduce the total delay to less than 0.75 switching period. In this paper, a robust damping method is proposed for a high-order filter-based single-phase grid-connected inverter, which can theoretically achieve the highest control bandwidth.

7.3 Proposed Single-Loop Control with High Bandwidth

7.3.1 Mathematical Model of the Proposed Single-Loop Control with Hybrid Damper

It is hard for the filter-based single-loop control to deal with the wide-range variation of the grid impedance [10]. Passive damping methods should be applied to achieve a wide bandwidth to reject the grid voltage harmonics effect. In [21], a

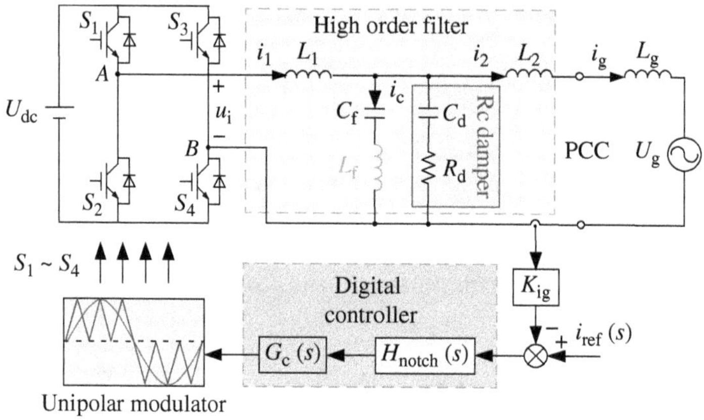

Figure 7.3 Single-phase grid-connected inverter with RC parallel damper and digital filter.

composite passive method was proposed for the *LLCL* filter to deal with the wide variation of the grid impedance at the cost of more power losses and material. In this paper, a hybrid single-loop damper is proposed not only to cut down the power losses but also to widen the control bandwidth.

Figure 7.3 shows a single-phase grid-connected inverter with an RC parallel damped *LLCL* filter, where the *PR* + *HC* regulator is adopted as the main controller. The block diagram is shown in Figure 7.4. The grid current i_2 is used as the feedback, and the transfer function of the grid current versus the output voltage of the inverter is then derived as:

$$G_{u_i \to i_g}(s) = \frac{Z_c(s)}{Z_1(s)Z_2(s) + Z_1(s)Z_c(s) + Z_2(s)Z_c(s)} \tag{7.10}$$

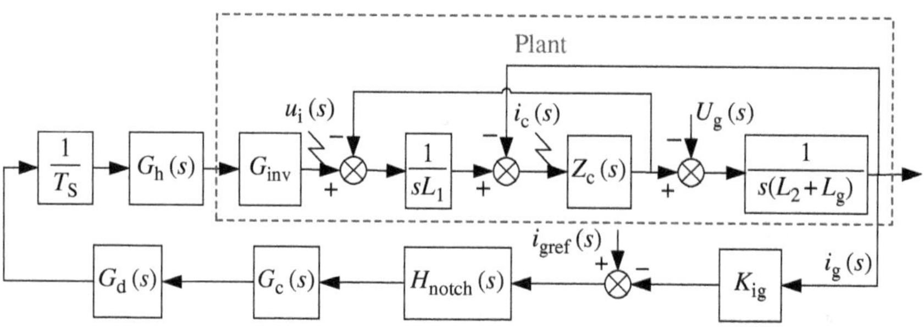

Figure 7.4 Block diagram of the single-phase grid-connected inverter with RC parallel damper and digital filter.

where,

$$Z_1(s) = L_1 s, Z_2(s) = (L_2 + L_g)s$$
$$Z_c(s) = \frac{(R_d C_d s + 1)(L_f C_f s^2 + 1)}{L_f C_f C_d s^3 + R_d C_d C_f s^2 + (C_d + C_f)s} \tag{7.11}$$

The open-loop transfer function of the system can be written as (7.12).

$$\begin{aligned} G_{\text{open-loop}}(s) &= \frac{i_g(s)}{i_{\text{ref}}(s)} \\ &= G_c(s) G_{\text{inv}} G_d(s) G_h(s) G_{u_i \to i_g}(s) K_{ig} H_{\text{notch}}(s) \\ &= G_{\text{plant}}(s) H_{\text{notch}}(s) \\ &= \frac{(R_d C_d s + 1)(L_f C_f s^2 + 1) G_c(s) G_{\text{inv}} K_{ig} H_{\text{notch}}(s) e^{-T_d s}}{s^3 L_1 (L_2 + L_g)[s^3 L_f C_f C_d + s R_d C_f C_d + (C_d + C_f)]} \\ &\quad + s(L_1 + L_2 + L_g)(R_d C_d s + 1)(L_f C_f s^2 + 1) \end{aligned} \tag{7.12}$$

where T_d is the total delay time, $H_{\text{notch}(s)}$ is the digital notch filter, and $G_{\text{plant}(s)} = G_c(s) G_{ui \to ig} G_{\text{inv}} K_{ig}$ is defined as the system plant. The transfer function of the *PR* + *HC* regulator is

$$G_c(s) = K_P + \sum_{h = 1,2,\cdots,n} \frac{k_{ih} s}{s^2 + (\omega_o (2h-1))^2} \tag{7.13}$$

where K_{ih} represents the individual resonant gain for the $(2h-1)$-order harmonic, and ω_0 is the grid angular frequency. After crossover frequency f_c, the *PR* controller can be simplified to K_p [17]. At the frequency where the phase of the system runs across −180°, the *LLCL* filter has almost the same characteristics as the *LCL* filter [1].When the digital filter is not taken into consideration, the gain margin and phase margin of the plant can be calculated with (7.14) and (7.15), respectively, where $\omega_r \sqrt{(L_1 + L_2 + L_g)/(L_1 (L_2 + L_g) C_f)}$ $\omega = 2\pi f$, f_{c_plant} is the crossover frequency of the plant, and f_{c180_plant} is the frequency when the phase of the plant crosses −180°.

To achieve a high control bandwidth to insert the harmonic compensators, the control delay is always expected to be as small as possible. Considering the general understanding and easy realization with the DSP digital controller, all the design is still based on $T_d = 0.75\,T_s$ [19].

$$\text{GM}_{\text{plant}} = -20 \log \left| \frac{(j\omega R_d C_d + 1) G_{\text{inv}} K_P K_{ig} / [L_1 (L_2 + L_g) C_f]}{\omega^2 R_d C_d (\omega^2 - \omega_r^2) - j\omega [\omega^2 (C_d / C_f + 1) + \omega_r^2]} \right|\Bigg|_{\omega = 2\pi f_{c180\_plant}} \tag{7.14}$$

$$\mathrm{PM_{plant}} = 180\circ + \varphi_{\mathrm{plant}}(j\omega)\big|_{\omega=2\pi f_{\mathrm{c\_plant}}}$$
$$= \arctan\omega R_{\mathrm{d}}C_{\mathrm{d}} + \arctan\frac{\omega(C_{\mathrm{d}}/C_{\mathrm{f}}+1)-4\pi^2 f_{\mathrm{r}}^2}{\omega R_{\mathrm{d}}C_{\mathrm{d}}(4\pi^2 f_{\mathrm{r}}^2-\omega^2)} - \frac{360}{2\pi}T_{\mathrm{d}}\omega\bigg|_{\omega=2\pi f_{\mathrm{c\_plant}}} \tag{7.15}$$

7.3.2 System-Characteristics-Based Single-Loop Control Design Methodology

The single-loop controller design includes the following six steps.

Step 1: Design of the RC Parallel Damper

The aim of the RC parallel damper is to reduce the Q-factor of the open-loop transfer function of the system at the dominant characteristic frequency. Generally, the larger capacitance, the better damping effect is achieved, but more additional reactive power is consumed, and more power losses are introduced [22]. To balance the damping effect and the additional power losses, C_{d} is often set to be the same as C_{f} as also shown in Figure 7.3.

$$\frac{\sqrt{\varepsilon+1}}{\varepsilon}\sqrt{\frac{L_{\mathrm{p}}+L_{\mathrm{f}}}{C_{\mathrm{f}}}} \le R_{\mathrm{d}} \le \frac{\varepsilon+1}{\varepsilon}\sqrt{\frac{L_{\mathrm{p}}+L_{\mathrm{f}}}{C_{\mathrm{f}}}} \tag{7.16}$$

where $\varepsilon = \frac{C_{\mathrm{d}}}{C_{\mathrm{f}}}$, $L_{\mathrm{p}} = \frac{L_1\cdot(L_2+L_{\mathrm{g}})}{L_1+L_2+L_{\mathrm{g}}}$. When $C_{\mathrm{d}} = C_{\mathrm{f}}$, the optimal value of the damping resistor can be calculated as,

$$R_{\mathrm{d}} = \frac{2+\sqrt{2}}{2}\sqrt{\frac{L_{\mathrm{p}}+L_{\mathrm{f}}}{C_{\mathrm{f}}}} \tag{7.17}$$

The resistor of the RC parallel damper is suggested to be designed under the weakest grid ($L_{\mathrm{g}} = L_{\mathrm{gmax}}$) when the gain margin and the power losses are taken into consideration.

Step 2: Design of the Proportionality Coefficient K_{p} of the *PR* + *HC* Regulator

Figure 7.5 shows the Bode diagrams of $G_{\mathrm{plant}\,(s)}$ with three different gains of the controller (*PR* + *HC* regulator). To maintain stability of the entire system, the minimum crossover frequency of the plant is set higher than the highest order harmonic compensator under the weakest grid condition. Then the minimum control gain can be calculated as:

$$K_{\mathrm{P\_min}} = \frac{2\pi f_{\mathrm{cmax\_w}}(L_1+L_2+L_{\mathrm{g}})}{G_{\mathrm{inv}}K_{\mathrm{ig}}} \tag{7.18}$$

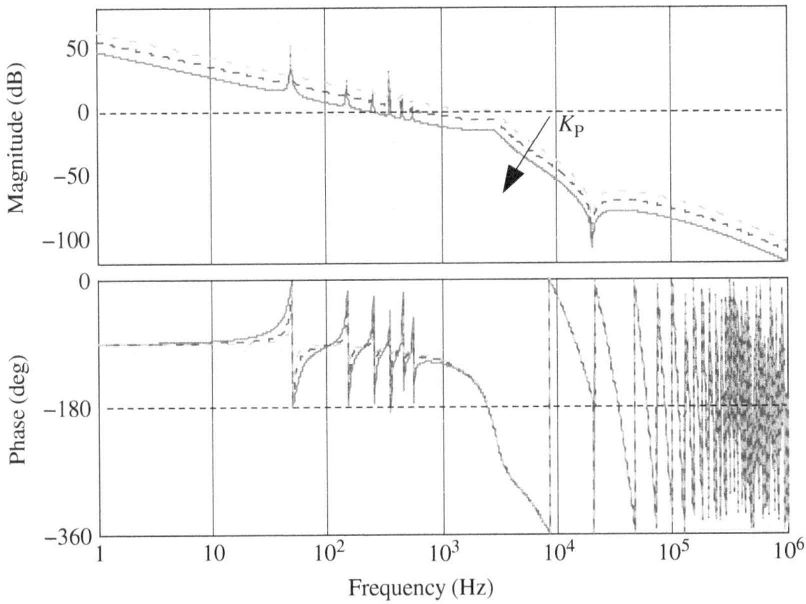

Figure 7.5 Bode diagram of $G_{plant}(s)$ with three different gains of the controller (*PR + HC* regulator).

At the same time, to ensure stability of the entire system under the stiffest grid condition, the control gain should be limited in the range as follows:

$$K_{P\_max} = \min\left(K_{P\_max\_1}, K_{P\_max\_2}\right) \tag{7.19}$$

where the $K_{P\_max\_1}$ and $K_{P\_max\_2}$ are the controller gains which are determined by $GM_{plant} = 0$ dB and $PM_{plant} = 45°$ (7.14) and (7.15) under the stiffest grid condition, respectively. If K_{P_max} is smaller than K_{P_min}, it is necessary to decrease the desired bandwidth until there is a proper value range.

Step 3: Determination of the Critical Grid Inductance

For the RC parallel damped system, the *Q*-factor at the dominant frequency varies with the grid inductance. Generally, the local maximum gain around the characteristic frequency will decrease when the equivalent grid impedance increases if $L_2 + L_g$ is large enough [21]. However, as shown in Figure 7.6, a local maximum gain around the characteristic frequency will appear for a certain grid inductance L_{g0}, when the value of L_2 is small. This critical grid inductance only depends on the parameters of the high-order filter and RC parallel damper.

The first-order partial derivative and the Hessian matrix of the gain of the plant can be derived using (7.20 and 7.21).

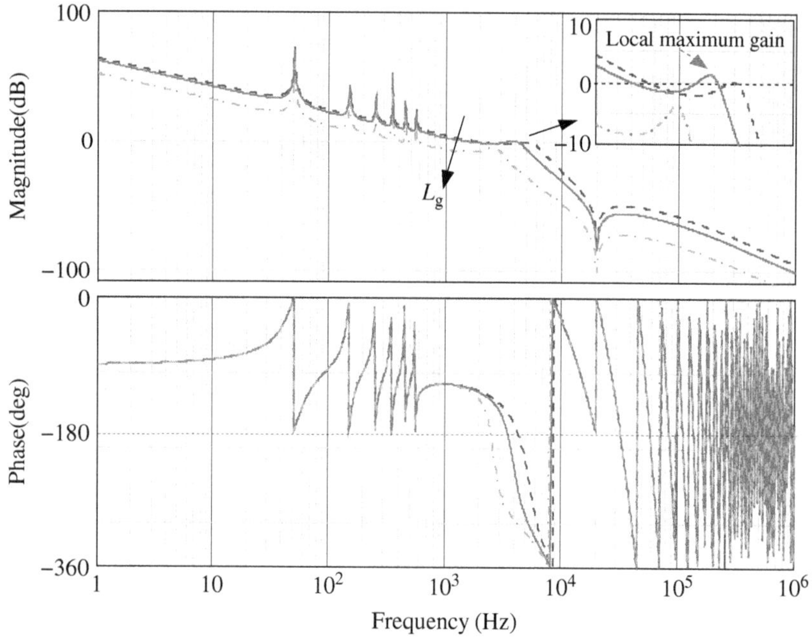

Figure 7.6 Bode diagrams of $G_{\text{plant}}(s, L_g)$ when the grid impedance varies in a wide range.

$$\begin{aligned} \frac{\partial |G_{\text{plant}}(j\omega, L_g)|}{\partial \omega} &= 0 \\ \frac{\partial |G_{\text{plant}}(j\omega, L_g)|}{\partial L_g} &= 0 \end{aligned} \tag{7.20}$$

$$H_f = \begin{vmatrix} \dfrac{\partial^2 |G_{\text{plant}}(j\omega,\ L_g)|}{\partial \omega^2} & \dfrac{\partial^2 |G_{\text{plant}}(j\omega,\ L_g)|}{\partial L_g \partial \omega} \\ \dfrac{\partial^2 |G_{\text{plant}}(j\omega,\ L_g)|}{\partial \omega \partial L_g} & \dfrac{\partial^2 |G_{\text{plant}}(j\omega,\ L_g)|}{\partial {L_g}^2} \end{vmatrix} \tag{7.21}$$

With (7.20 and 7.21), the critical grid inductance L_{g0} and the related frequency f_0, where the phase of the system with the RC parallel damper runs across −180° can be calculated. Then they can be used to design the digital filter. The relationship between GM_{plant} and the grid inductance is shown in Figure 7.7. As shown in Figure 7.7, two types of the relationship between the grid inductance L_g and GM_{plant} will appear. For Case 1, the minimum gain margin of the plant occurs at the critical inductance L_{g0}. A digital filter should be designed to offer enough gain attenuation, when the grid inductance varies between the area of $[L_{gl}, L_{gu}]$, and at least β dB gain attenuation at the critical grid inductance L_{g0}. Case 2 will occur if L_2 is large enough, where the minimum GM_{plant} appears under

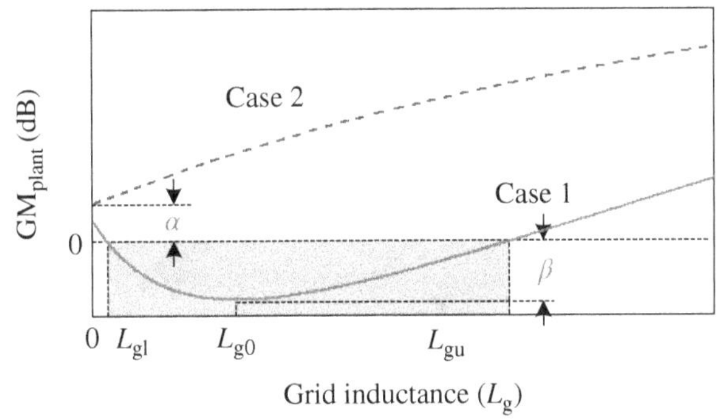

Figure 7.7 Relationship between grid inductance L_g and gain margin of GM_{plant}.

the stiffest grid condition. In Case 2, if α is larger than 3 dB after a proper K_P is selected according to the desired bandwidth of the plant, the digital filter is not necessary. Otherwise, a digital filter should also be inserted in the stiffest grid state to improve the system performance.

Step 4: Determination of the Critical Frequency Region for Case 1 and the Critical Frequency (f_0 of Case 1 and f_{L0} of Case 2)

The design of the digital filter is based on the frequency region. Therefore, it is necessary to determine the frequency corresponding to the critical grid inductance obtained in Step 3. The phase of the plant is a function of the frequency and the grid impedance, so the relationship between the grid inductance and f_{c180_plant} can be derived by (7.22) and it is shown in Figure 7.8a.

$$\varphi_{plant}(s,\ L_g)\big|_{s=j2\pi f} = -180° \tag{7.22}$$

As shown in Figure 7.8a and Figure 7.7, f_{c_plant} is a monotone decreasing function of the grid inductance, the critical frequency f_0 of Case 1 is corresponding to the critical inductance L_{g0}. The critical frequency f_{L0} of Case 2 is corresponding to zero grid inductance of the stiffest grid (in the landing mode).

Similarly, the relationship between the GM_{plant} and f_{c_plant} can be obtained by (7.23) and it is plotted in Figure 7.8b.

$$GM_{plant} = -G_{plant}(s,\ L_g)\big|_{s=j2\pi f_{c180\_plant}} \tag{7.23}$$

From Figure 7.8, it can be seen that for a given plant of Case 1, when the equivalent grid impedance is L_{g0}, if no other active damping is adopted, the minimum GM_{plant} will appear at the frequency of f_0 and in the frequency range of $[f_L, f_U]$,

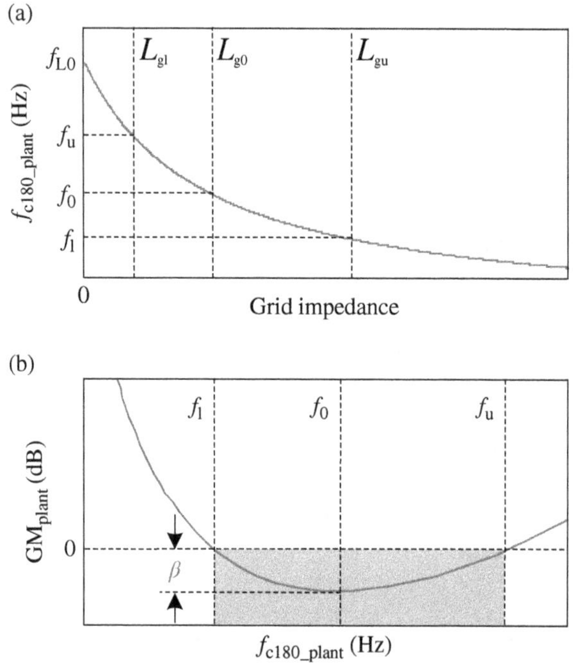

Figure 7.8 (a) The relationship between the grid impedance L_g and f_{c180_plant}. (b) The relationship between the frequency when the phase of the plant crosses −180° f_{c180_plant} and the gain margin of the plant GM_{plant}.

where the loop gain of β dB should be compensated. While for a given plant of Case 2, if no other active damping is adopted, the minimum GM_{plant} will appear at the frequency of f_{LO} in the island mode.

Step 5: Design of the Digital Notch Filter

According to Figure 7.7 and Figure 7.8, there may be many kinds of digital filters that can meet the requirement [9]. To achieve a good phase margin of the entire system, the notch filter is adopted in this paper. The transfer function of the notch filter is shown in (7.24).

$$H_{notch}(s) = \frac{s^2 + 2\xi_1\omega_{nr}s + \omega_{nr}{}^2}{s^2 + 2\xi_2\omega_{nr}s + \omega_{nr}{}^2} \tag{7.24}$$

where the Bode diagram is depicted in Figure 7.9. It can be found that the notch filter gets the maximum depth M = 20log (ξ_1/ξ_2) at the center frequency $f_{nr} = \omega_{nr}/2\pi$, which can be calculated with (7.25 and 7.26).

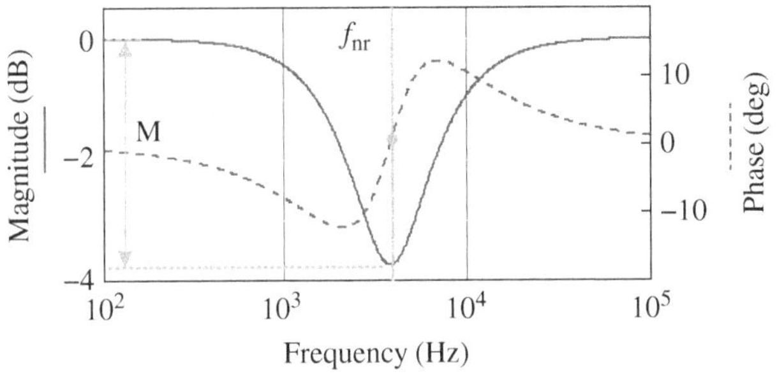

Figure 7.9 Bode diagram of the designed digital notch filter.

$$|H_{\text{notch}}(j\omega)| = 20\log\left(\sqrt{\frac{(\omega_{\text{nr}}{}^2-\omega^2)^2+(2\xi_1\omega_{\text{nr}}\omega)^2}{(\omega_{\text{nr}}{}^2-\omega^2)^2+(2\xi_2\omega_{\text{nr}}\omega)^2}}\right) \tag{7.25}$$

$$\varphi_{\text{notch}}(j\omega) = \arctan\left(\frac{2\xi_1\omega_{\text{nr}}\omega}{\omega_{\text{nr}}{}^2-\omega^2}\right) - \arctan\left(\frac{2\xi_2\omega_{\text{nr}}\omega}{\omega_{\text{nr}}{}^2-\omega^2}\right) \tag{7.26}$$

Therefore, (7.27 and 7.28) are applied to determine the notch filter to obtain a 3 dB gain margin for the Case 1 and the Case 2, respectively

$$\begin{cases} f_{\text{nr}} = f_0 \\ M = 20\log(\xi_1/\xi_2) \geq \beta + 3 \end{cases} \tag{7.27}$$

$$\begin{cases} f_{\text{nr}} = f_{L_0} \\ M = 20\log(\xi_1/\xi_2) \geq 3 - \alpha \end{cases} \tag{7.28}$$

By considering the phase effect of the notch filter and the dynamic performance, after choosing a proper value of ξ_2, it is necessary to decrease ξ_1 until the gain margin of the entire system is always satisfactory, when the grid impedance varies within a wide range.

Step 6: Checking the Phase Margin of the Entire System

After the above five steps, it is necessary to check the phase margin of the entire system using

$$\begin{aligned}\text{PM} &= 180^\circ + \varphi_{\text{plant}}(j\omega) + \varphi_{\text{nocth}}(j\omega)\big|_{\omega=2\pi f_c} \\ &= \arctan\omega R_{\text{d}}C_{\text{d}} + \arctan\frac{\omega^2(C_{\text{d}}/C_{\text{f}}+1)-4\pi^2 f_{\text{r}}^2}{\omega R_{\text{d}}C_{\text{d}}(4\pi^2 f_{\text{r}}^2-\omega^2)} + \varphi_{\text{nocth}}(j\omega) - \frac{360}{2\pi}T_{\text{d}}\omega\bigg|_{\omega=2\pi f_c}\end{aligned} \tag{7.29}$$

If the phase margin of the entire system is not always larger than 45° owing to the digital filter when the grid inductance varies within a wide range, then the designer should go back to Step 2 to reduce K_P until the phase margin of the entire system is satisfactorily met.

Figure 7.10 shows algorithm summarizing the different steps of the proposed hybrid damper design methodology for high-order filter-based grid-connected inverter.

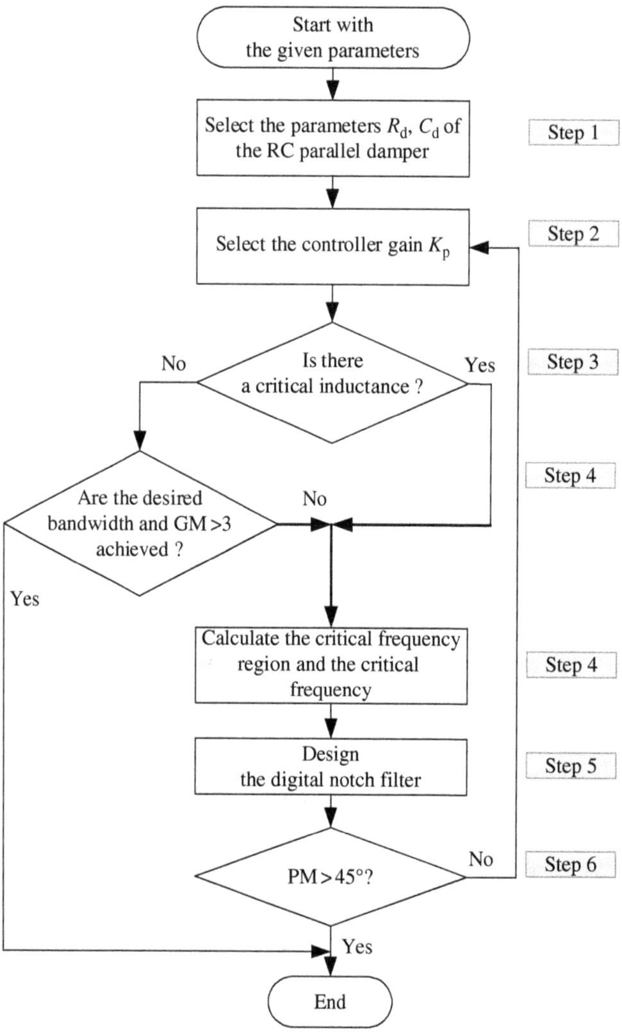

Figure 7.10 Design procedure for the hybrid damper.

7.4 Design Example

7.4.1 System Design

To illustrate the design of the hybrid damper for a high-order filter-based grid-connected inverter, an example of a 2 kW, 220 V/50 Hz *LLCL* filter-based system is shown stepwise according to the methodology introduced in Section III, where the parameters are shown as $f_s = 20$ kHz, $U_{dc} = 350$ V, $U_{tri} = 0.25$ V, $K_{ig} = 0.0182$, $L_1 = 1.2$ mH, $L_2 = 0.22$ mH, $L_f = 32$ μH, $C_f = 2$ μF, and an 11th-order harmonic compensator is desired.

The resistor of the RC parallel damper is chosen as 35 Ω according to (7.17) and C_d is equal to C_f to balance the damping performance and the additional power losses.

1) Under the stiffest grid condition, if $GM_{plant} = 0$, K_{P_max1} can be calculated as 0.946 using (7.14) and if $PM_{plant} = 45°$, K_{P_max2} can be calculated as 0.845 using (7.15). Under the weakest grid condition, to avoid the phase effect of the *HC* regulator, if the minimum cross-over frequency of the plant is set to 580 Hz, K_{P_min} can be calculated as 0.749 using (7.18). Finally, the controller gain of K_P is selected as 0.76, while K_{ih} is then set to 100.
2) With the selected $K_P = 0.76$, the relationship between the grid inductance and GM_{plant} can be drawn according to (7.14), which is similar to the Case 1 as shown in Figure 7.8, and the inductance region which makes the system unstable is (0.21, 1.12 mH). The critical grid inductance L_{g0} is 0.54 mH and $\beta = 0.556$ dB.
3) It could be found that the frequency range corresponding to the critical grid inductance region is (3292, 4446 Hz), and the critical frequency is 3790 Hz, which are calculated using (7.22 and 7.23).
4) To preserve a gain margin of 3 dB, the transfer function of the notch filter can be achieved from (7.27). After checking the gain margin, the notch filter is designed as,

$$H_{notch}(s) = \frac{1.76 \cdot 10^{-9}s^2 + 4.37 \cdot 10^{-5}s + 1}{1.76 \cdot 10^{-9}s^2 + 6.72 \cdot 10^{-5}s + 1} \tag{7.30}$$

where $\omega_{nr} = 7580\pi$, $\xi_1 = 0.52$, $\xi_2 = 0.8$.

Based on bilinear transformation, when the sampling frequency is 20 kHz, the transfer function of the digital filter in the discrete domain can be derived as,

$$H_{notch}(z) = \frac{0.855 - 0.56z^{-1} + 0.319z^{-2}}{1 - 0.56z^{-1} + 0.174z^{-2}} \tag{7.31}$$

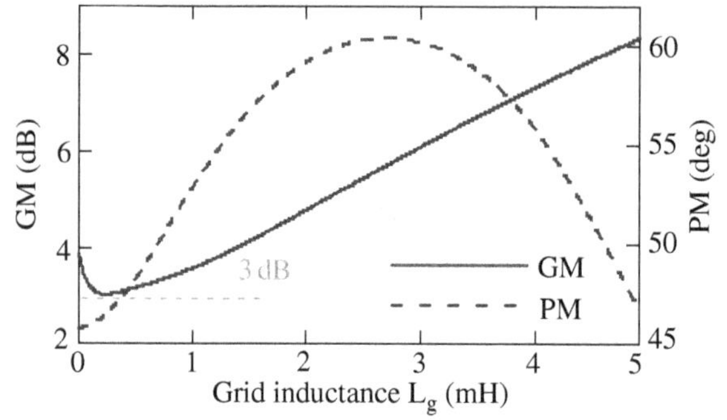

Figure 7.11 Gain and phase margins of the entire system with RC parallel damper and digital notch filter.

5) Figure 7.11 shows the gain margin and the phase margin of the entire system with the proposed hybrid damper. It can be seen that the gain margin is always larger than 3 dB and the minimum phase margin is larger than 45°, which can meet the requirement for the design specification.

7.4.2 System Parameter Robustness Analysis

In Figure 7.12, the designed system stability margin is presented when all the filter parameters vary 30% to verify the system robustness. It is seen in Figure 7.12 that the designed system can keep stable when the filter parameters and the grid impedance vary within a wide range, which verifies the system robustness.

7.5 Experimental Verification

Based on the theoretical analysis, a prototype of the single-phase full-bridge grid-connected inverter with the DSP (TMS320LF2812A) controller is constructed. The discontinuous unipolar PWM modulation method is adopted. An external inductor is used to emulate the equivalent grid inductance. All the parameters are the same as those for the theoretical analysis in Section 7.6.

A programmable ac source (Chroma 6530) is used to emulate the grid voltage distorted by third, fifth, seventh, ninth harmonics. The magnitudes of harmonics with respect to the grid fundamental voltage are 9.38%, 3.44%, 0%, and 1.15%, respectively. The measured grid voltages and currents of the *LLCL* filter-based inverter system are shown in Figure 7.13.

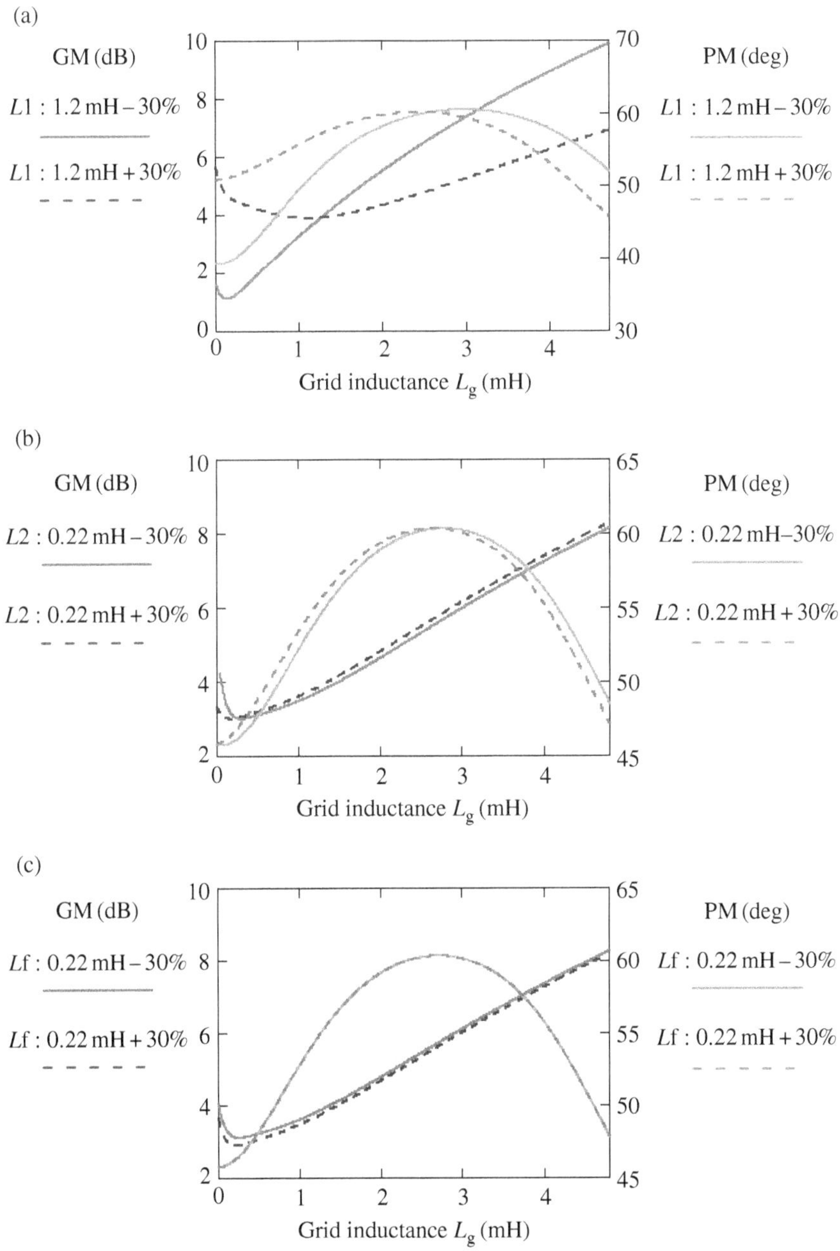

Figure 7.12 Stability margin of the design system when the filter parameters vary (a) L_1 varies ±30%; (b) L_2 varies ±30%; (c) L_f varies ±30%.

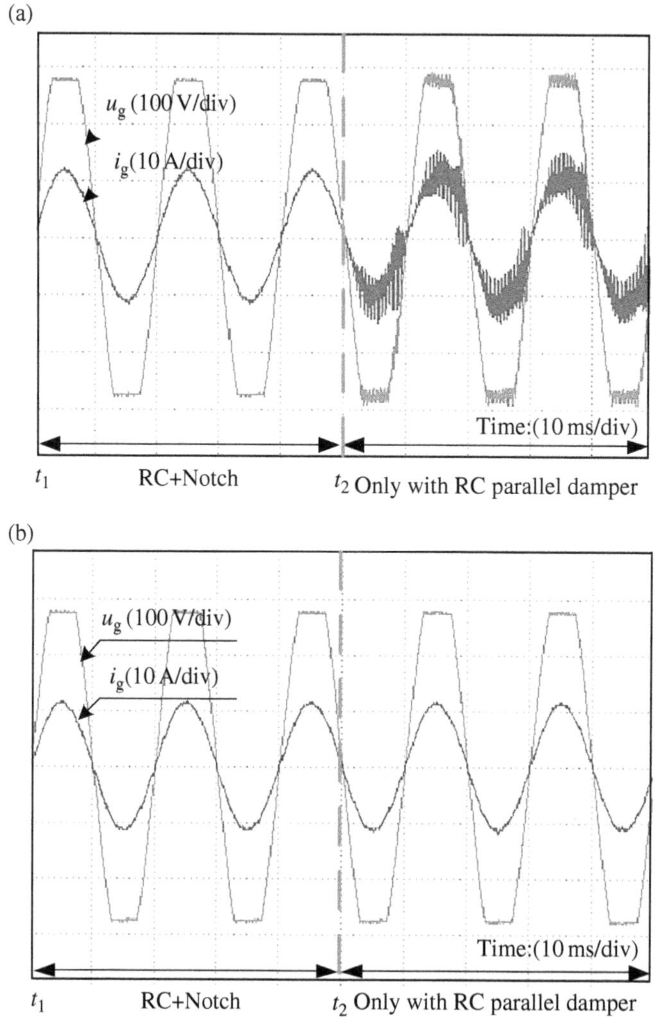

Figure 7.13 Measured grid voltages and currents of the *LLCL* filter-based system with the digital filter and the RC parallel damper when the harmonics are injected into the grid voltage and the digital filter is cut off at t_2 (a) under the critical grid (L_g = 0.54 mH), (b) under the weakest grid (L_g = 4.8 mH).

As shown in Figure 7.13a, under the critical grid condition, only the proposed hybrid-damper-based system can be maintained stable, while the traditional RC parallel damped system cannot. Moreover, as shown in Figure 7.13b, it can be seen that in the weakest grid state, both the conventional RC parallel damping and the proposed hybrid damping methods can maintain the *LLCL* filter-based grid-connected inverter system stable when the grid voltage is distorted by significant harmonics.

When the grid voltage is distorted by 3rd, 5th, 7th, 9th, 11th harmonics, whose magnitudes with respect to the grid fundamental voltage are 1.2%, 2.8%, 1.3%, 2.4%, and 1.5%, respectively, and the corresponding phases are 180°, 0°, 180°, 0°, and 180°, the grid voltage U_g and the current injected into the grid i_g in the weakest grid is shown in Figure 7.14a. Figure 7.14b is the FFT of U_g and i_g. It can be found out that the system can be kept stable and can suppress up to the 11th harmonic well.

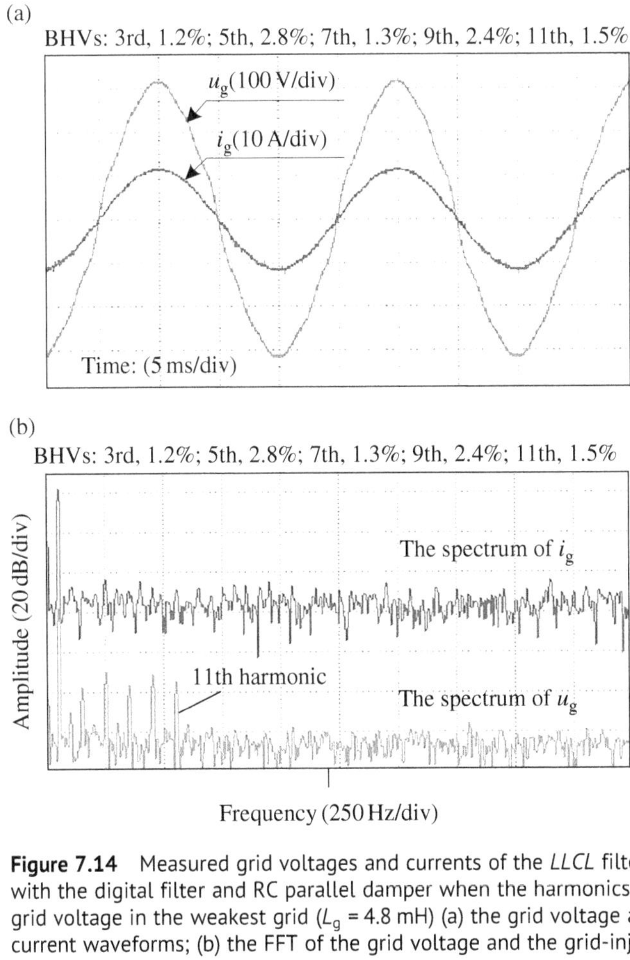

Figure 7.14 Measured grid voltages and currents of the *LLCL* filter-based system with the digital filter and RC parallel damper when the harmonics are injected into the grid voltage in the weakest grid (L_g = 4.8 mH) (a) the grid voltage and the grid-injected current waveforms; (b) the FFT of the grid voltage and the grid-injected current.

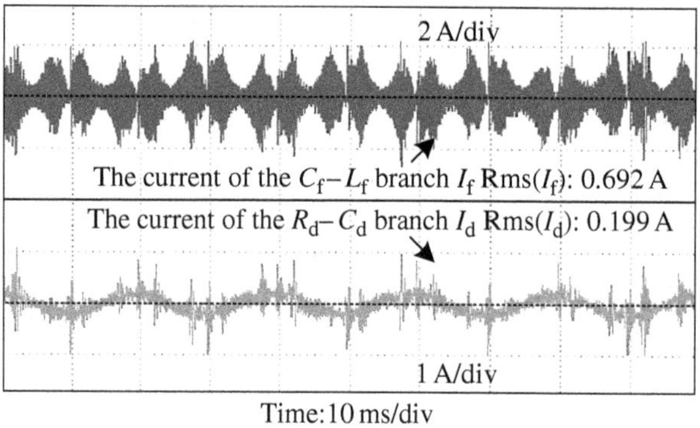

Figure 7.15 Measured currents of the C_f–L_f branch and the C_d–R_d branch of the *LLCL* filter-based system with the digital filter and the RC parallel damper in the weakest grid.

Figure 7.15 shows the measured currents (If , Id) of the $Cf - Lf$ branch and $Rd - Cd$ branch of the *LLCL*-filter-based system with the digital filter and the *RC* parallel damper in the weakest grid state. It can be calculated that the power losses at the *RC* parallel damper can be limited to $(I^2 dRd/Po)\% = 0.07\%$, where Po is the power injected into the grid. Experimental results are in good agreement with the theoretical analysis.

7.6 Summary

In this chapter, different AD methods have been first described. Moreover, a single-loop hybrid damping method is proposed. A stepped design procedure is described in detail. Afterward, following conclusions can be drawn,

A small grid-side inductor (L_2) can reduce the total inductance for a high-order filter-based grid-connected inverter, where the *LLCL* filter is preferred due to the requirement of the IEEE 1547.2-2008 standard.

A small value of L_2 will result in the control difficulties of the grid-connected inverter, when the grid voltage has a large amount of background harmonics.

To resist the effect caused by the BHV, the control bandwidth of *LCL* or *LLCL* filter-based grid-connected inverter is expected as high as possible.

When the equivalent grid impedance varies within a wide range, only a single-loop AD design for the small grid-side inductor (L_2)-based grid-connected inverter can hardly meet the bandwidth requirement to suppress the effect of the BHV.

For the *LLCL* filter-based grid-connected inverter, the proposed hybrid damping method can almost achieve the highest bandwidth under the same delay and grid conditions, especially when the total delay is reduced to less than 0.75 switching period.

Due to the RC parallel damper, the proposed hybrid damping method brings extra power losses, which are so small that this can be ignored for the *LLCL* filter-based single-phase grid-connected inverter.

Experiments on a 2 kW *LLCL* filter-based single-phase grid-connected inverter prototype with a fixed controller gain fully verify the effectiveness of the proposed damping method, when the grid inductance varies from 0 to 4.8 mH, where the weak grid coefficient is $\eta = 4.4$.

References

1 Wu, W., He, Y., and Blaabjerg, F. (2012). An LLCL power filter for single-phase grid-tied inverter. *IEEE Trans. Power Electron.* 27 (2): 782–789.

2 Xu, J., Xie, S., and Tang, T. (2014). Improved control strategy with grid-voltage feedforward for LCL-filter-based inverter connected to weak grid. *IET Power Electron.* 7 (10): 2660–2671.

3 Li, W., Ruan, X., Pan, D., and Wang, X. (2013). Full-feedforward schemes of grid voltages for a three-phase LCL-type grid-connected inverter. *IEEE Trans. Ind. Electron.* 60 (6): 2237–2250.

4 Yang, D., Ruan, X., and Wu, H. (2014). Impedance shaping of the grid-connected inverter with LCL filter to improve its adaptability to the weak grid condition. *IEEE Trans. Power Electron.* 29 (11): 5795–5805.

5 Wang, X., Blaabjerg, F., and Loh, P.C. (2015). Virtual RC damping of LCL-filtered voltage source converters with extended selective harmonic compensation. *IEEE Trans. Power Electron.* 30 (9): 4726–4737.

6 Shen, G., Zhu, X., Zhang, J., and Xu, D. (2010). A new feedback method for PR current control of LCL-filter-based grid-connected inverter. *IEEE Trans. Ind. Electron.* 57 (6): 2033–2041.

7 Timbus, A., Liserre, M., Teodorescu, R. et al. (2009). Evaluation of current controller for distributed power generation systems. *IEEE Trans. Power Electron.* 24 (3): 654–664.

8 Liserre, M., Teodorescu, R., and Blaabjerg, F. (2006). Stability of photovoltaic and wind turbine grid-connected inverters for a large set of grid impedance values. *IEEE Trans. Power Electron.* 21 (1): 263–272.

9 Dannehl, J., Liserre, M., and Fuchs, F.W. (2011). Filter-based active damping of voltage source converters with LCL filter. *IEEE Trans. Ind. Electron.* 58 (8): 3623–3633.

10 Alzola, R.P., Liserre, M., Blaabjerg, F. et al. (2014). A self-commissioning notch filter for active damping in a three-phase LCL -filter-based grid-tie converter. *IEEE Trans. Power Electron.* 29 (12): 6754–6761.

11 Pan, D., Ruan, X., Bao, C. et al. (2014). Capacitor-current-feedback active damping with reduced computation delay for improving robustness of LCL-type grid-connected inverter. *IEEE Trans. Power Electron.* 29 (7): 3414–3427.

12 Alzola, R.P., Liserre, M., Blaabjerg, F. et al. (2014). Systematic design of the lead-lag network method for active damping in LCL-filter based three-phase converters. *IEEE Trans. Ind. Inf.* 10 (1): 43–52.

13 Pan, D., Ruan, X., Bao, C. et al. (2015). Optimized controller design for LCL-type grid-connected inverter to achieve high robustness against grid-impedance variation. *IEEE Trans. Ind. Electron.* 62 (3): 1537–1547.

14 Xu, J., Xie, S., and Tang, T. (2014). Active damping-based control for grid-connected LCL-filtered inverter with injected grid current feedback only. *IEEE Trans. Ind. Electron.* 61 (9): 4746–4758.

15 Liu, F., Zhou, Y., Duan, S. et al. (2009). Parameter design of a two-current-loop controller used in a grid-connected inverter system with LCL filter. *IEEE Trans. Ind. Electron.* 56 (11): 4483–4491.

16 Yang, D., Ruan, X., and Wu, H. (2015). A real-time computation method with dual sampling modes to improve the current control performances of the LCL-type grid-connected inverter. *IEEE Trans. Ind. Electron.* 62 (7): 4563–4572.

17 Bao, C., Ruan, X., Wang, X. et al. (2014). Step-by-step controller design for LCL-type grid-connected inverter with capacitor-current-feedback active-damping. *IEEE Trans. Power Electron.* 29 (3): 1239–1253.

18 Eren, S., Pahlevaninezhad, M., Bakhshai, A., and Jain, P.K. (2013). Composite nonlinear feedback control and stability analysis of a grid-connected voltage source inverter with LCL filter. *IEEE Trans. Ind. Electron.* 60 (11): 5059–5074.

19 Grogan, S.A.S., Holmes, D.G., and McGrath, B.P. (2011). High-performance voltage regulation of current source inverters. *IEEE Trans. Power Electron.* 26 (9): 2439–2448.

20 Holmes, D.G., Lipo, T.A., McGrath, B.P., and Kong, W.Y. (2009). Optimized design of stationary frame three phase AC current regulators. *IEEE Trans. Power Electron.* 24 (11): 2417–2426.

21 Wu, W., Sun, Y., Huang, M. et al. (2014). A robust passive damping method for LLCL-filter-based grid-tied inverters to minimize the effect of grid harmonic voltages. *IEEE Trans. Power Electron.* 29 (7): 3279–3289.

22 Channegowda, P. and John, V. (2010). Filter optimization for grid interactive voltage source inverters. *IEEE Trans. Ind. Electron.* 57 (12): 4106–4114.

8

Robust Impedance-Based Design of *LLCL*-Filtered Grid-Connected Inverter against the Wide Variation of Grid Reactance

8.1 Introduction

During the design of grid-connected inverters, the effects of the equivalent grid reactance, including both inductive and capacitive impedances, on the stability of the system must be addressed [1–7]. However, when the capacitive grid impedance is solved, the detailed parameter design method of the grid-connected inverter based on the *LCL* or *LLCL* filter has not been introduced yet. In this chapter, passivity-based analysis is extended to the switching frequency, in which two non-passive regions are identified. Once the output admittance of the inverter intersects with the grid admittance in the two non-passive regions, the entire system becomes unstable. Based on the analysis, considering the wide variation of the grid reactance, to ensure stability of the entire system, an exact and robust system parameter design method is proposed. Using this parameter design method, the entire system can operate stably under a wide variation in the grid reactance.

In this chapter, we first model the *LLCL*-type grid-connected inverter. Then, the stability of the grid-connected inverter based on the *LLCL* filter under a wide range of grid reactances is analyzed. Next, two basic solutions are recommended: to minimize the non-passive area and avoid intersections located in the non-passive region. Finally, to verify the feasibility of the proposed measures, simulations and experiments were performed on a 2 kW/220 V *LLCL*-filter-based single-phase grid-connected inverter prototype. The experimental results show that the proposed parameter design method can achieve a high-performance grid-connected inverter system under rigid grid conditions.

Control and Filter Design of Single-Phase Grid-Connected Converters, First Edition. Weimin Wu, Frede Blaabjerg, Henry Chung, Yuanbin He, and Min Huang.

Published 2023 by John Wiley & Sons, Inc.

8.2 Modeling of the *LLCL*-Type Grid-Connected Inverter

8.2.1 System Description

Figure 8.1 shows a single *LLCL*-filter-based grid-connected inverter connected to the ideal grid via the equivalent reactances of C_g and L_g. (L_1, L_2, C_f, L_f) and (L_g, C_g) are parameters of the filter and grid reactance, respectively.

The plant-equivalent block diagram is shown in Figure 8.2, together with impedances Z_1, Z_{LC}, and Z_2 for L_1, L_f C_f trap, and L_2, respectively. G_c (s) is the current controller, G_d (s) is the total delay time, and $G_{inv} = U_{dc}/U_{tri}$ is the gain of the inverter, where U_{dc} is the dc voltage and U_{tri} is the amplitude value of the triangular carrier.

The proportional resonant (PR) and harmonic compensator (HC) controllers are adopted as the current controllers in this study, while the transfer function of G_c (s) is given as follows:

$$G_c(s) = K_p + \sum_{h=1,3,5,7,9,\cdots} \frac{K_{ih}s}{s^2 + (\omega_o h)^2} \tag{8.1}$$

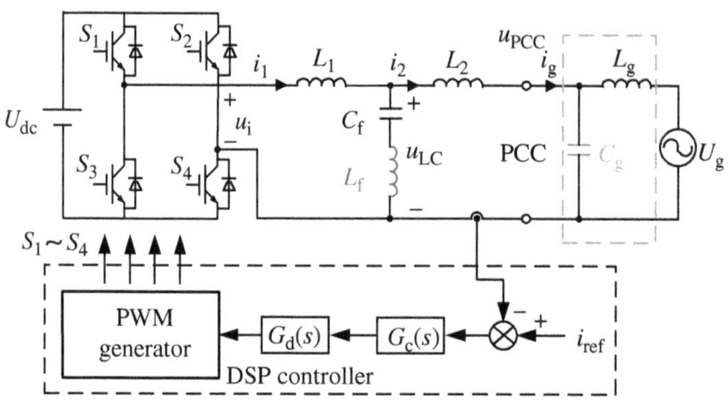

Figure 8.1 Single *LLCL*-filter-based grid-connected inverter connected to the ideal grid via the equivalent reactances of C_g and L_g.

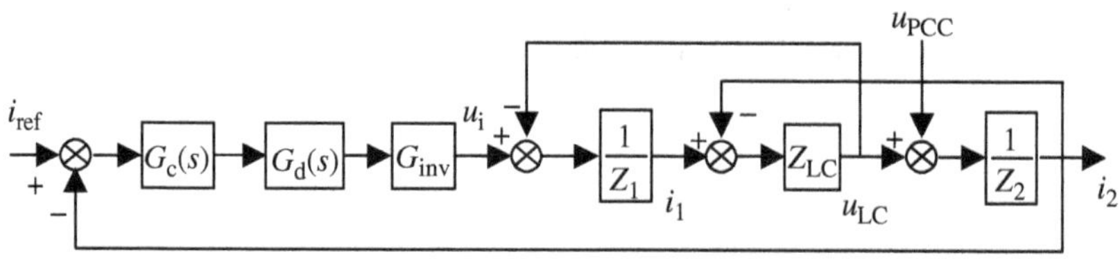

Figure 8.2 Block diagram of the grid-connected inverter with a single-loop grid-injected current control.

where ω_o ($=2\pi f_o$) is the fundamental angular frequency, and K_p and K_{ih} are the proportional and resonant gains, respectively.

The total delay time $G_d(s)$, including the computational delay and modulation delay, is commonly expressed as follows:

$$G_d(s) = \frac{\sin \omega T_s/2}{\omega T_s/2} e^{-\lambda T_s s} \tag{8.2}$$

where $T_s = 1/f_s$ is the sampling period and λ is the delay time normalized with T_s. The normal value of λ is selected as 1 or 1.5 in a real operation [8].

8.2.2 Norton Equivalent Model

Figure 8.3 shows the Norton equivalent model of the *LLCL*-filter-based grid-connected inverter with the grid-injected current feedback control, where the inverter output admittance can be obtained as follows:

$$Y_O(s) = \frac{s^2 C_f L_f + 1}{s^3 C_f (L_1 L_2 + L_1 L_f + L_2 L_f) + s^2 K_p G_{inv} C_f L_f \frac{\sin \omega T_s/2}{\omega T_s/2} e^{-\lambda T_g s} + s(L_1 + L_2) + K_p G_{inv} \frac{\sin \omega T_s/2}{\omega T_s/2} e^{-\lambda T_g s}} \tag{8.3}$$

In addition, the grid admittance and its resonance frequency can be, respectively, expressed as follows:

$$Y_g(s) = \frac{s^2 L_g C_g + 1}{s L_g} \tag{8.4}$$

$$f_g = \frac{1}{2\pi\sqrt{C_g L_g}} \tag{8.5}$$

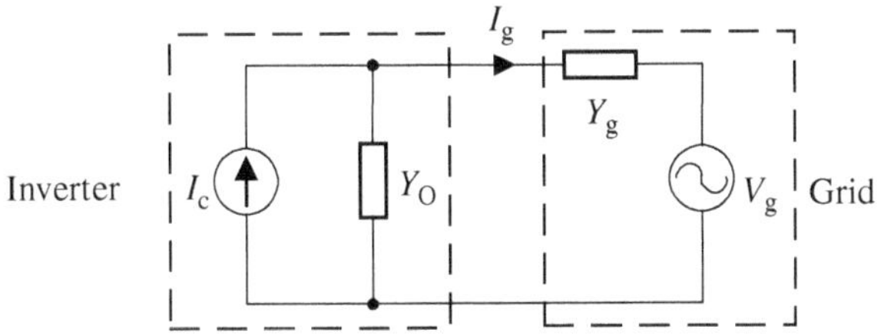

Figure 8.3 Norton equivalent model of the *LLCL*-filter-based grid-connected inverter with the grid-injected current feedback control.

8.3 Stability Analysis Considering Grid-Reactance Variation

8.3.1 Non-Passive Regions of Inverter Output Admittance

To be passive, a linear continuous system $G(s)$ must satisfy the following two requirements at frequency ω [9]:

1) $G(s)$ has no right-half-plane (RHP) poles, and
2) $\text{Re}\,[G\,(j\omega)] \geq 0 \leftrightarrow \arg\,[G\,(j\omega)] \in [-90°, 90°]\ \forall\ \omega > 0$.

Then, the passivity violation can be identified by inspecting the negative real part of the inverter output admittance. The real part of (8.3) is calculated as follows:

$$\text{Re}\,[Y_{\text{O}}(j\omega)] = \frac{K_{\text{P}}G_{\text{inv}}[1 - C_{\text{f}}(L_1 + L_{\text{f}})\omega^2](1 - C_{\text{f}}L_{\text{f}}\omega^2)\dfrac{\sin \omega T_{\text{s}}/2}{\omega T_{\text{s}}/2}\cos(\lambda T_{\text{s}}\omega)}{A^2 + B^2} \tag{8.6}$$

where A and B are defined as follows:

$$A = K_{\text{p}}G_{\text{inv}}\left(1 - C_{\text{f}}L_{\text{f}}\omega^2\right)\frac{\sin \omega T_{\text{s}}/2}{\omega T_{\text{s}}/2}\cos(\lambda T_{\text{s}}\omega)$$

$$B = (L_1 + L_2)\omega - C_{\text{f}}(L_1L_2 + L_1L_{\text{f}} + L_2L_{\text{f}})\omega^3 - K_{\text{p}}G_{\text{inv}}\left(1 - C_{\text{f}}L_{\text{f}}\omega^2\right)\frac{\sin \omega T_{\text{s}}/2}{\omega T_{\text{s}}/2}\sin(\lambda T_{\text{s}}\omega)$$

Within the switching frequency of ω_{s}, the term $(1-C_{\text{f}}L_{\text{f}}\omega^2)$ will always be positive. Meanwhile, the denominator of (8.6) is always positive. Note that K_{p}, G_{inv}, and $\sin(\omega T_{\text{s}}/2)/(\omega T_{\text{s}}/2)$ are also always positive; therefore, the polarity of $\text{Re}\,[Y_{\text{o}}(j\omega)]$ depends on the term $[1-C_{\text{f}}(L_1 + L_{\text{f}})]\omega^2$ and the cosine term.

When $[1-C_{\text{f}}(L_1 + L_{\text{f}})]\omega^2 = 0$, the resonant frequency f_{p} can be calculated as follows:

$$f_{\text{p}} = \frac{1}{2\pi\sqrt{C_{\text{f}}(L_1 + L_{\text{f}})}} \tag{8.7}$$

For the cosine term, f_{d1} and f_{d2} are both critical frequencies, which are defined in (8.8).

$$\begin{cases} f_{\text{d1}} = \dfrac{f_s}{4\lambda} \\ f_{\text{d2}} = \dfrac{3f_s}{4\lambda} \end{cases} \tag{8.8}$$

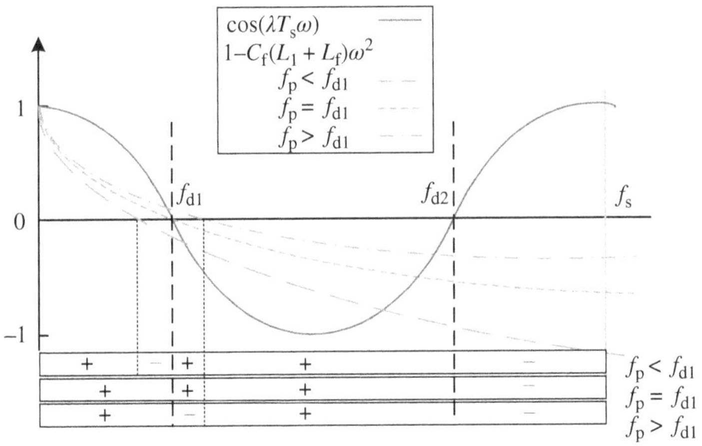

Figure 8.4 Non-passive regions of inverter output admittance.

The plots of cos $(\lambda T_s\omega)$ and $[1-C_f(L_1 + L_f)]\omega^2$, together with their combined polarities, are shown in Figure 8.4, where the term $[1-C_f(L_1 + L_f)]\omega^2$ is drawn for three frequency cases of f_p. The region $[f_{d2}, f_s]$ is always non-passive. When f_p is smaller than f_{d1}, a non-passive region $[f_p, f_{d1}]$ appears. The non-passive region $[f_{d1}, f_p]$ also appears when f_p is larger than f_{d1}. Note that within f_{d2}, as long as f_{d1} is equal to f_p, the first non-passive region disappears.

According to the above analysis, the non-passive regions with two different cases (Case A: $f_p < f_{d1}$, Case B: $f_p > f_{d1}$) are summarized in Figure 8.5. The non-passive region $[f_p, f_{d1}]$ or $[f_p, f_{d1}]$ is defined as the non-passive region I (NPR_I), while $[f_{d2}, f_s]$ is defined as non-passive region II (NPR_II).

8.3.2 Possible Instability Under the Wide Variation of Grid Reactance

In fact, the real distributed grid inductance and capacitance are unknown and may vary over a wide range, which causes a wide variation in f_g. Therefore, the intersection of the grid admittance and the inverter output admittance may also vary over a wide range and are located in NPR_I or NPR_II. Once the absolute value of the phase difference at the intersection becomes greater than 180°, the system becomes unstable according to the impedance-based stability criterion [10–12]. Taking Case A as an

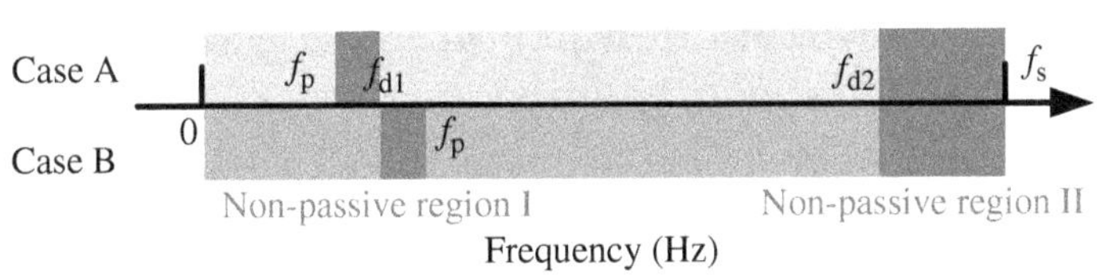

Figure 8.5 Summary of non-passive regions.

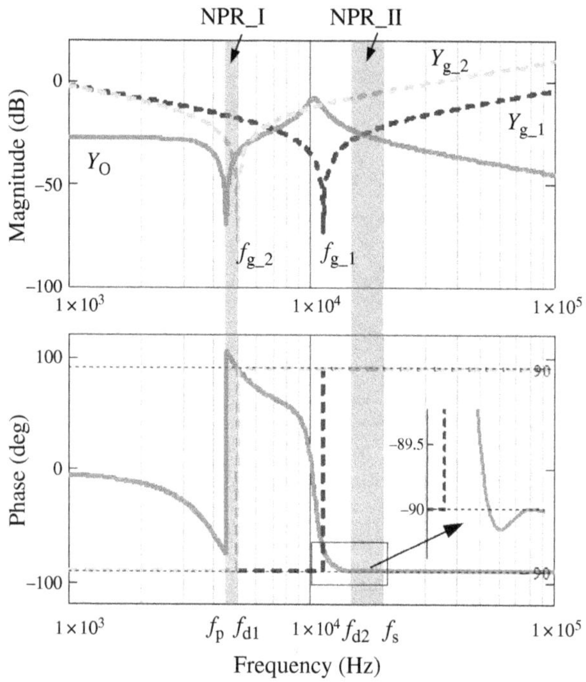

Figure 8.6 Bode plots of the inverter output admittance and the grid admittance with two different f_g values.

example, Figure 8.6 shows the Bode plots of the inverter output admittance and the grid admittance with two different f_g values, where the system appears unstable.

Based on the analysis above, we find that to stabilize the grid-connected inverter under a wide variation of grid reactance, the following two basic strategies can be employed:

1) Minimize the non-passive regions as much as possible, which is preferentially recommended.
2) Avoid the intersection of the inverter output admittance and grid admittance falling in the non-passive regions of the grid-connected inverter.

8.4 Proposed Measures and Design Procedure Under the Grid-Reactance Variation Condition

8.4.1 Proposed Measures Against Grid-Reactance Variation

As shown in Figure 8.5, if $f_p = f_{d1}$, NPR_I can be eliminated. However, it cannot always be guaranteed because f_p depends on parameters (L_1, C_f, L_f) of the filter, which may shift in the actual application.

Furthermore, as shown in Figure 8.6, when f_g is sufficiently high, the intersection of the grid admittance and the inverter output admittance will possibly fall in NPR_II. The minimum value of L_gC_g should be set to avoid the intersection falling into NPR_II. In fact, to improve the electromagnetic interference (EMI) noise suppression performance of the high-order-filter-based grid-connected inverter, a parallel capacitance C_{EMI} is generally required [13–15], which can guarantee a lower limit value of grid capacitance C_g.

Therefore, measures to enhance the stability of the presented grid-connected inverter system against the grid-reactance variation are proposed as follows:

1) According to the regulation of the resonant parameters based on (8.7), we set $f_p = f_{d1}$ to eliminate NPR_I.
2) Select the minimum capacitance C_{g_min} to limit the intersection of the grid reactance and the inverter output admittance under the stiffest grid condition, so that the intersection will never fall in NPR_II.
3) Add an RC damper to reshape the grid admittance and ensure system stability when the original intersection is located in NPR_I owing to parameter drifting of the filter.

According to the above measures, Figure 8.7 shows the structure of the proposed filter for the grid-connected inverter, where C_{EMI}, R_d, and C_d are the EMI capacitors of C_{EMI}, the damping resistor, and capacitor, respectively.

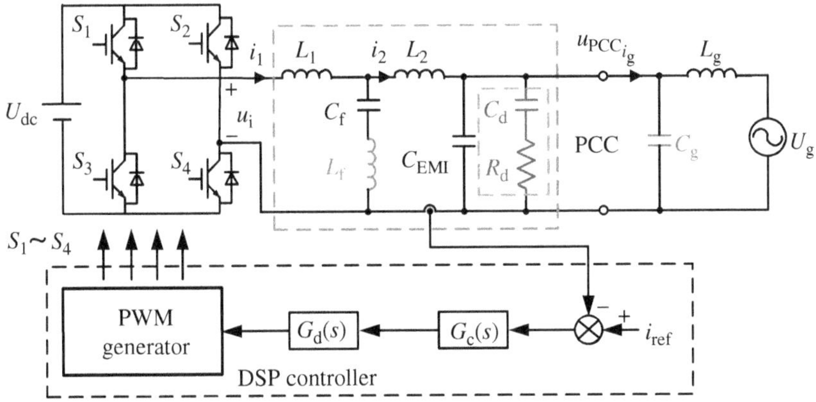

Figure 8.7 Proposed filter structure for the grid-connected inverter.

8.4.2 Design Procedure

The single-loop current control design contains the following seven steps, and the design flowchart is shown in Figure 8.8.

Step 1- Calculate the Minimum Grid Inductance L_{g_min}

The real grid inductance consists of the transformer leakage inductance, L_{leak}, and the power line inductance, L_{line}. Considering x as the short-circuit inductance, L_{leak} can be calculated as follows:

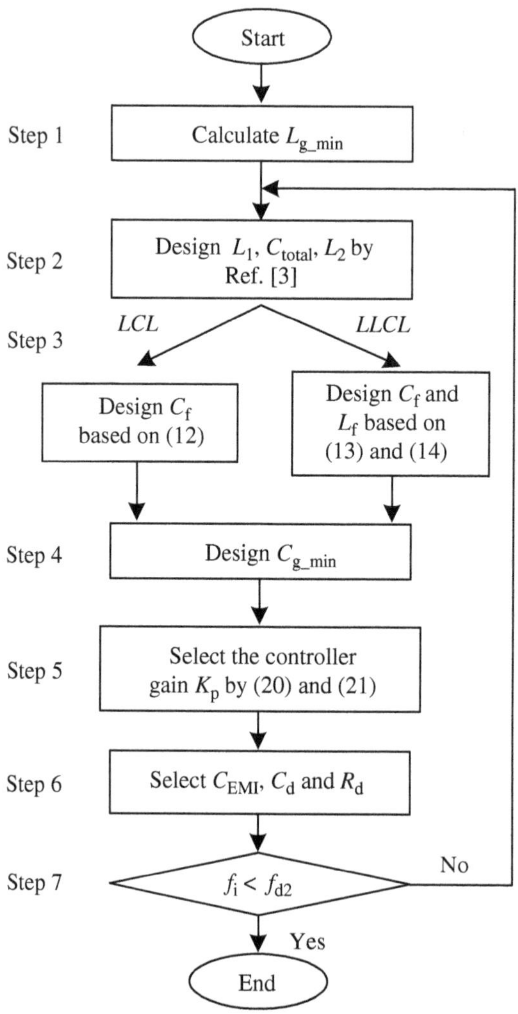

Figure 8.8 Flowchart of the proposed design.

$$L_{\text{leak}} = x\frac{U_{\text{g}}^2}{\omega_{\text{o}}P} \tag{8.9}$$

where x is generally between 4 and 7%, U_{g} is the grid fundamental root-mean-square voltage, ω_{o} is the fundamental frequency in radians per second, and P is the rated capacity of the transformer.

Considering the minimum grid inductance, L_{line} is neglected, $L_{\text{g\_min}} = L_{\text{leak}}$.

Step 2- Design L_1, C_{total}, and L_2

The *LCL*- or *LLCL*-filter designs were introduced in [16], and the design principles for inductors L_1 and L_2 were not changed.

1) Select the tolerable current ripple on the converter side to the design of the converter-side inductor L_1.

$$15\% \le \frac{\Delta I_1}{I_{\text{ref}}} = \frac{1}{8}\frac{2U_{\text{dc}}}{L_1 f_{\text{s}} I_{\text{ref}}} \le 40\% \tag{8.10}$$

   where ΔI_1 is the converter-side current ripple, and I_{ref} is the rated reference peak current.

2) Select the reactive power absorbed in the rated condition to the total capacitance C_{total}. Considering y as the percentage of reactive power absorbed in the rated condition,

$$C_{\text{total}} = C_{\text{f}} + C_{\text{g\_min}} \le \frac{5\% P_{\text{rated}}}{U_{\text{g}}^2 \omega_{\text{o}}} \tag{8.11}$$

   where P_{rated} is the rated generated power.

3) Note that if L_{f} of the *LLCL* filter is set to zero, then the traditional *LCL* filter will be obtained. For an *LCL* filter, L_2 mainly depends on the objective of attenuating each harmonic around the switching frequency down to 0.3%, which is in accordance with the IEEE Standard 519-2014.

For an *LLCL* filter, owing to the L_{f}–C_{f} trap, the current harmonics around the switching frequency satisfy the requirements of IEEE Standard 519-2014 with far more ease. Therefore, L_2 is designed to attenuate harmonics around the second integer multiple of the switching frequency to be less than 0.3%.

Step 3- Design the Bypass Filtering Branch

For an *LCL* filter, the capacitor C_{f_LCL} can be calculated using (8.12) according to $f_p = f_{d1}$ ($L_f = 0$).

$$C_{f\_LCL} = \frac{16\lambda^2}{L_1(2\pi f_s)^2} \tag{8.12}$$

For an *LLCL*-filter, the resonant capacitor C_{f_LLCL} can be calculated using (8.13) according to $f_p = f_{d1}$. Thus, the resonant inductor L_f can be selected at the converter switching frequency f_s according to (8.14). However, the series trap is also influenced by its quality factor Q expressed in (8.15), where R_f represents the gapped equivalent resistance of inductor L_f. The value of Q should be considered between the usual range of $10 \leq Q \leq 50$ [16].

$$C_{f\_LLCL} = \frac{16\lambda^2 - 1}{L_1(2\pi f_s)^2} \tag{8.13}$$

$$L_f = \frac{1}{C_{f\_LLCL}(2\pi f_s)^2} \tag{8.14}$$

$$Q = \frac{1}{R_f}\sqrt{\frac{L_f}{C_f}} \tag{8.15}$$

Step 4- Design the Minimum Grid Capacitance C_{g_min}

A very small value of C_g may occur in the actual power grid, resulting in the mentioned intersection falling in NPR_II, as shown in Figure 8.5. Therefore, the minimum grid capacitance C_{g_min} should be chosen to guarantee the lower limit value of the grid capacitance, so that the variation range of the grid reactance is reduced. Then, C_{g_min} is selected as follows:

$$C_{g\_min} = C_{total} - C_f \tag{8.16}$$

Step 5- Design the Proportional Gain K_P of the *PR+HC* Regulator

From Figure 8.1, the open-loop transfer function of the proposed system can be written as follows:

$$\begin{aligned} G_{ol}(s) &= G_c(s)G_d(s)G_{inv}G_{filter}(s) \\ &= \frac{1}{C_gL_gC_fL_f}\frac{\left(s^2+\omega_g^2\right)\left(s^2+\omega_f^2\right)}{s\left(as^4+bs^2+c\right)}G_c(s)G_d(s)G_{inv} \end{aligned} \tag{8.17}$$

where $G_{filter}(s)$ is the transfer function of the grid-injected current i_2 versus the output voltage u_i of the inverter, and a, b, and c are defined as follows:

$$a = C_g L_g \left(C_f L_1 L_2 + C_f L_f L_1 + C_g L_f L_2 \right)$$
$$b = C_f L_1 \left(L_2 + L_f + L_g \right) + C_f L_f \left(L_2 + L_g \right) + C_g L_g \left(L_1 + L_2 \right)$$
$$c = L_1 + L_2 + L_g$$

The current controller $G_c(s)$ is usually simplified as K_P because the resonance terms have a negligible influence on the system stability if the maximum control bandwidth is well set. The gain margin and phase margin of the system can then be calculated as follows:

$$\mathrm{GM} = -20 \log \left| K_P G_{\mathrm{inv}} G_{\mathrm{filter}}(j\omega) \right| \Big|_{\omega = 2\pi f_{d1}} \tag{8.18}$$

$$\mathrm{PM} = 180^\circ + \left(\angle G_{\mathrm{filter}}(j\omega) - \frac{360}{2\pi} \lambda T_s \omega \right) \Bigg|_{\omega = 2\pi f_c} \tag{8.19}$$

where f_c is the crossover frequency of the system.

The minimum crossover frequency of the system is generally set higher than the highest harmonic compensator frequency under the weakest grid condition of the maximum equivalent grid inductance and maximum real grid capacitance. Then, the minimum control gain can be calculated as follows:

$$K_{P\_min} = \frac{1}{G_{\mathrm{inv}} G_{\mathrm{filter}}(j\omega)} \Bigg|_{\omega = 2\pi f_{cweak}} \tag{8.20}$$

where f_{cweak} is the minimum crossover frequency.

Simultaneously, to ensure a sufficient stability margin for the entire system, the maximum control gain should be chosen using the following equation:

$$K_{P\_max} = \min\left(K_{P\_stiff1}, K_{P\_stiff2} \right) \tag{8.21}$$

where K_{P_stiff1} and K_{P_stiff2} are the controller gains, which are determined by GM = 3 dB and PM = 30° according to (8.18) and (8.19), respectively, under the stiffest grid condition of real grid inductance and minimum real capacitance C_{g_min}, as shown in Figure 8.9. If K_{P_max} is smaller than K_{P_min}, it is necessary to reduce the desired control bandwidth until there is a proper value range.

Step 6- Select C_{EMI}, C_d, and R_d

Although f_p is equal to f_{d1}, NPR_I still appears when the filter parameters drift. Once the intersection is located in NPR_I and the absolute value of the phase difference is beyond 180°, the entire system becomes unstable. To guarantee the stability of the entire system, a novel RC damper designed to reshape the grid admittance is first proposed, which is shown in Figure 8.7, aiming at limiting the phase difference of the inverter output admittance Y_o and the damped grid admittance Y_{grc} between −180°and 180°.

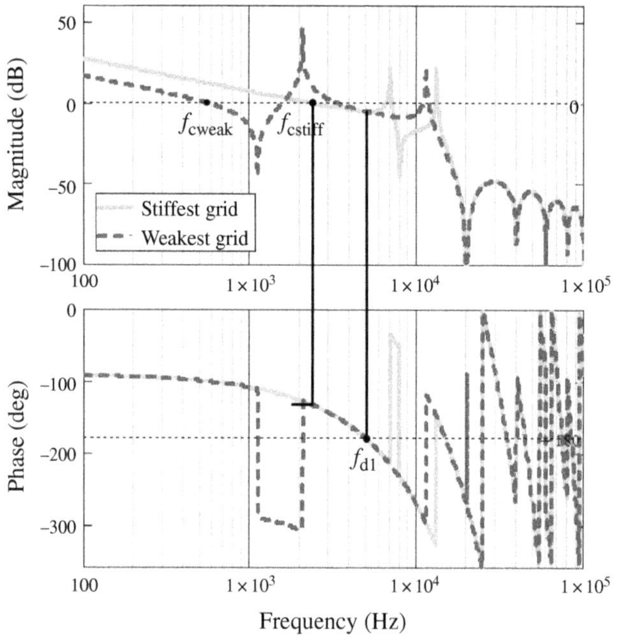

Figure 8.9 Bode plots of the system under the stiffest grid condition with C_{g_min} and the weakest grid condition with the maximum grid capacitance.

The damping capacitor C_d can be split from C_{g_min}, and the rest of C_{g_min} is used as the C_{EMI}. To balance the damping effect and damping losses, an equal value of C_{EMI} and C_d may be a suitable selection [17]. Then, C_{EMI} and C_d can be selected as follows:

$$C_{EMI} = C_d = \frac{C_{g\_min}}{2} \tag{8.22}$$

Regardless of $[f_p, f_{d1}]$ or $[f_{d1}, f_p]$, the phase difference of the inverter output admittance and grid admittance is the largest when f_g is close to f_p. The RC damper can increase the intersection frequency and reduce the phase difference. Therefore, the damping resistor value is selected by the equivalent Q-factor method [18], which can be represented as follows:

$$Q_E = \frac{1}{R_E}\sqrt{\frac{L_E}{C_E}} \tag{8.23}$$

where Q_E is the equivalent Q-factor, and R_E, L_E, and C_E are the equivalent resistance, inductance, and capacitance of the equivalent series LCR circuit, respectively.

As shown in Figure 8.7, the equivalent resistance R_E and capacitance C_E of the grid capacitance, C_{EMI}, and RC parallel damper can be calculated as follows:

$$\begin{aligned}
&\frac{1}{s\left(C_{EMI}+C_g\right)}//\left(\frac{1}{sC_d}+R_d\right)\Bigg|_{s=j\omega}\\
&=\frac{R_dC_ds+1}{R_dC_d\left(C_{EMI}+C_g\right)s^2+\left(C_d+C_{EMI}+C_g\right)s}\Bigg|_{s=j\omega}\\
&=R_E+\frac{1}{sC_E}
\end{aligned} \tag{8.24}$$

Then, at the dominant resonance frequency, the equivalent resistance R_E, inductance L_E, and capacitance C_E of the equivalent LCR circuit can be calculated as follows:

$$\begin{aligned}
R_E&=\frac{R_d\,C_d^2}{\left(C_d+C_{EMI}+C_g\right)^2-\left(R_dC_d\left(C_{EMI}+C_g\right)s\right)^2}\Bigg|_{s=j2\pi f_{res}}\\
L_E&=L_g\\
C_E&=\frac{\left(R_dC_d\left(C_{EMI}+C_g\right)s\right)^2-\left(C_d+C_{EMI}+C_g\right)^2}{R_d^2\,C_d^2\left(C_{EMI}+C_g\right)s^2-\left(C_d+C_{EMI}+C_g\right)}\Bigg|_{s=j2\pi f_{res}}
\end{aligned} \tag{8.25}$$

Substituting (8.25) into (8.23), the Q-factor at the dominant resonance frequency can be calculated. Therefore, the Q-factor as a function of the damping resistor and the grid inductance and capacitance are plotted in Figure 8.10.

As shown in Figure 8.10, the Q-factor increases while R_d decreases and C_g and L_g increase. Therefore, R_d is selected to determine the optimal Q-factor under the condition $f_g = f_p$. Note that in the reasonable range, the larger the value of R_d, the greater is the damping power loss.

The damped grid admittance is expressed as follows:

$$Y_{grc}(s)=\frac{s^3L_gC_gR_dC_d+s^2L_g\left(C_g+C_d\right)+sR_dC_d+1}{s^2L_gR_dC_d+sL_g} \tag{8.26}$$

Finally, the Bode plots of the inverter output admittance and the damped grid admittance are plotted to verify whether the phase difference remains below 180° when $f_g = f_p$.

Step 7- Check $f_i < f_{d2}$

After the aforementioned six steps, whether the maximum frequency f_i of the intersection is located in NPR_II under the stiffest grid condition with C_{g_min} must be verified, as shown in Figure 8.11. Directly solving the exact value of f_i is

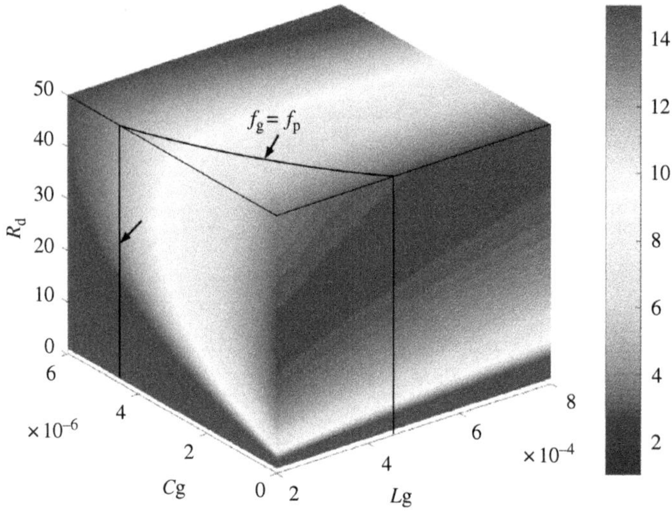

Figure 8.10 Relationship between the *Q*-factor, damper resistor, grid inductance, and grid capacitance for the damped grid admittance.

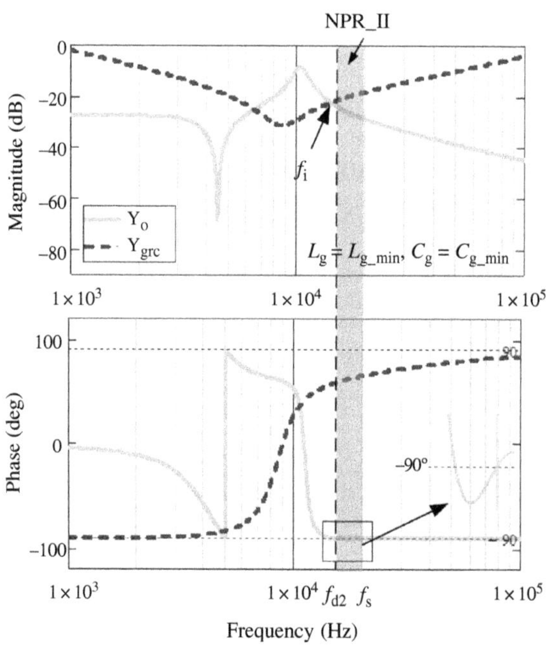

Figure 8.11 Bode plots of the inverter output admittance and the damped grid admittance under the stiffest grid condition with C_{g_min}.

challenging. Therefore, a simple engineering method is developed. Under the conditions of $L_g = L_{g\_min}$ and $C_g = C_{g\_min}$, if the magnitude of the inverter output admittance is smaller than that of the damped grid admittance at f_{d2}, then f_i is smaller than f_{d2}, and the system is stable. The magnitude difference of the inverter output admittance and the damped grid admittance are defined as follows:

$$\Delta Y(f,\ L_g,\ C_g)\Big|_{\substack{L_g = L_{g\_min},\\ C_g = C_{g\_min},\\ f = f_{d2}}} = Y_{Oa}(f) - Y_{grca}(f, L_g, C_g) \tag{8.27}$$

where Y_{Oa} (f) and Y_{grca} $(f,\ L_g,\ C_g)$ are the magnitudes of $Y_O(s)$ and $Y_{grc}(s)$, respectively.

If $\Delta Y(f,\ L_g,\ C_g)\Big|_{\substack{L_g = L_{g\_min},\\ C_g = C_{g\_min},\\ f = f_{d2}}}$ is smaller than zero, then f_i is not in NPR_II. Subsequently, the design procedure was completed. Otherwise, *Step 2* is returned to the reselected parameters.

8.5 Design Example

Assume that there is a 40 kW system that consists of 20 sets of 2 kW 220 V/50 Hz inverters. To illustrate the proposed parameter design for the grid-connected inverter under the wide variation of grid reactance, an example of a 2 kW 220 V/50 Hz *LLCL*-filter-based system is designed step by step according to the procedure introduced in Section 8.4, where the parameters are shown as $f_s = 20$ kHz, $U_{dc} = 350$ V, and $U_{tri} = 0.25$ V, and an 11th harmonic compensator is desired.

The systematic design procedure is shown as follows:

1) $P = 40$ kW, $x = 5.2\%$, L_{leak} is calculated as 0.2 mH according to (8.9). Then, L_{g_min} is equal to 0.2 mH.
2) The 30% current ripple is adopted for the converter-side inductor L_1, and L_1 is selected to be 1.2 mH. The total capacitor value C_{total} is designed as 2.8 μF to limit the reactive power to 2.36%. If some of the constraints cannot be met, it should be increased to a limit of 5%. L_2 is designed as 0.22 mH to attenuate harmonics around the second integer multiple of the switching frequency to be less than 0.3%.
3) For $\lambda = 1$, the resonant capacitor C_f can be calculated as 0.8 μF using (8.13). The resonant inductor L_f was selected as 80 μH using (8.14). The series resistor $R_f = 0.2\ \Omega$ and the quality factor Q is calculated as 50 by (8.15).
4) C_{g_min} is selected as 2 μF according to (8.16).

5) Considering the maximum real grid capacitance $C_g = 60\ \mu F$, the real grid inductance is 0.2 mH under the stiffest grid condition, whereas the equivalent grid capacitance and inductance become 3 μF and 4 mH under the weakest grid condition, respectively, if there are 20 inverters in parallel. If the minimum crossover frequency of the system f_{cweak} is set to 550 Hz, K_{P_min} can be calculated as 0.016 using (8.20). Under the stiffest grid condition with C_{g_min}, if GM = 3 dB, K_{P_stiff1} can be calculated as 0.019 using (8.19), and if PM = 30°, K_{P_stiff2} can be calculated as 0.022 using (8.18). Finally, the controller gain of K_P is selected as 0.017, whereas K_{ih} is set to 18.2.
6) C_{EMI} and C_d are both chosen to be 1 μF according to (8.22). When C_f drifts +5%, L_1 and L_f drift +25%, the value of f_p becomes 4.34 kHz. To ensure a Q-factor less than 12 when $L_g = 0.2$ mH and $C_g = 4.5\ \mu F$, R_d is selected as 25 Ω according to Figure 8.10. To verify whether the phase difference remains below 180° near f_p, the Bode plots of the inverter output admittance and the damped grid admittance are shown in Figure 8.12. The damped grid admittance does not interact with the inverter output admittance in NPR_I.

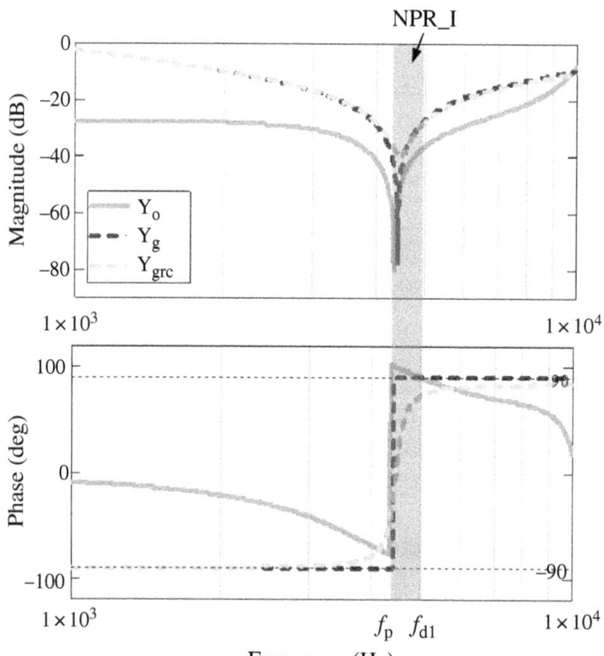

Figure 8.12 Bode plots of the inverter output admittance and the damped grid admittance when $L_g = 0.2$ mH and $C_g = 4.5\ \mu F$.

Table 8.1 Designed parameters of the converter.

Parameter	Symbol	Value
Inverter-side inductor	L_1	1.2 mH
Parasitic resistance	R_{L1}	0.1 Ω
Resonant inductor	L_f	80 μH
Series resistance	R_f	0.2 Ω
Resonant capacitor	C_f	0.8 μF
Grid-side inductor	L_2	0.22 mH
Parasitic resistance	R_{L2}	0.01 Ω
Gain of inverter	G_{inv}	1400
Proportional gain	K_P	0.017
Resonant gain	K_{ih}	18.2
Delay time coefficient	Λ	1
Switching frequency	f_s	20 kHz
Resonance frequency	f_p	5 kHz
Minimum grid capacitor	C_{g_min}	2 μF
Critical frequency	f_{d1}	5 kHz
	f_{d2}	15 kHz
Minimum crossover frequency	f_{cweak}	550 Hz
Capacitor for suppressing the EMI noise	C_{EMI}	1 μF
Grid impedance shaping resistor	R_d	25 Ω
Grid impedance shaping capacitor	C_d	1 μF

7) The value of $\Delta Y(f,\ L_g,\ C_g)\big|_{\substack{L_g = L_{g\_min},\\ C_g = C_{g\_min},\\ f = f_{d2}}}$ is negative. Therefore, the design procedure ends.

Finally, the designed parameters of the converter are listed in Table 8.1.

8.6 Simulation and Experimental Verification

8.6.1 Simulation

Simulations were carried out with the PSIM software to confirm the effectiveness of the proposed measures. The designed converter parameters are shown in Table 8.1, and the calculated non-passive region NPR_II is [19, 20] kHz. The four

Table 8.2 Grid parameters.

Condition	Normal operation		Parameter drifting	
Parameter	Case 1	Case 2	Case 3	Case 4
L_g	0.3 mH	0.3 mH	0.51 mH	0.51 mH
R_g	0.06 Ω	0.06 Ω	0.1 Ω	0.1 Ω
C_g	1 μF	1 μF	1 μF	1 μF
C_{EMI}		1 μF	1 μF	1 μF
R_d				25 Ω
C_d				1 μF

contradistinctive cases of the grid parameters used for the simulation are also shown in Table 8.2. Among them, Case 1 and Case 2 are designed under normal operation, whereas Cases 3 and 4 are designed when the filter parameters drift.

For Case 1, the intersection frequency is calculated as 15.6 kHz, which is located in NPR_II. Furthermore, the simulated waveform and FFT waveform of the grid-injected current are shown in Figures 8.13 and 8.14, where the excessive grid-injected current harmonics appear around the frequencies of 15.6 kHz and 4.4 kHz, respectively.

Based on Case 1, Case 2 uses an extra capacitor C_{EMI} to avoid the intersection from falling into NPR_II. The simulated grid-injected current in Figure 8.15 shows that the system operates stably again.

However, $f_p = f_{d1}$ cannot always be guaranteed because the filter parameters may drift in actual applications. In our case, C_f and L_f drifts +25% and −20%,

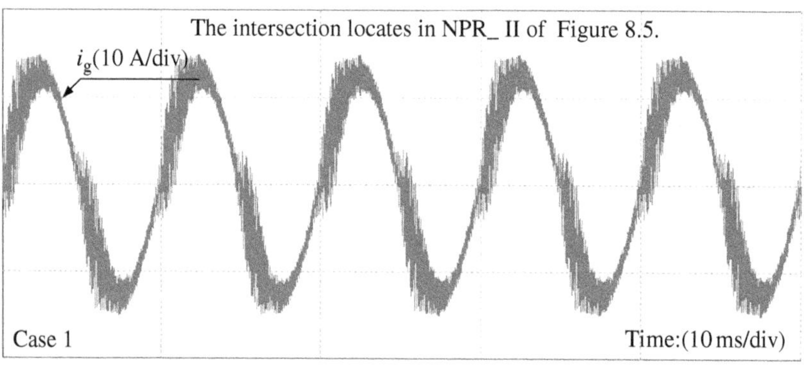

Figure 8.13 Simulated grid-injected current of the *LLCL*-filter-based system with parameters of Case 1, when the intersection locates in NPR_II of Figure 8.5.

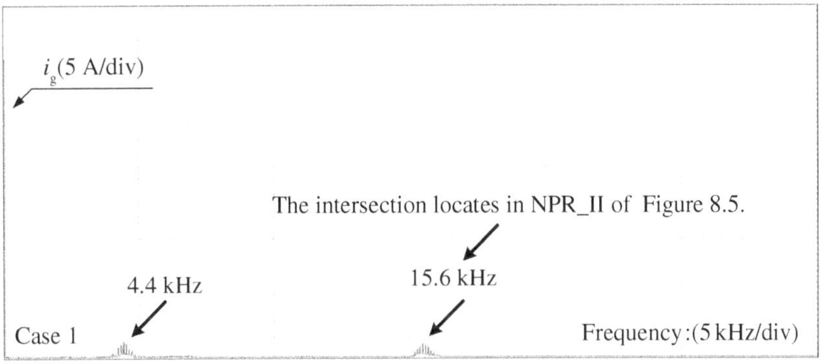

Figure 8.14 Simulated FFT waveform of the grid-injected current with parameters of Case 1, when the intersection is located in NPR_II of Figure 8.5.

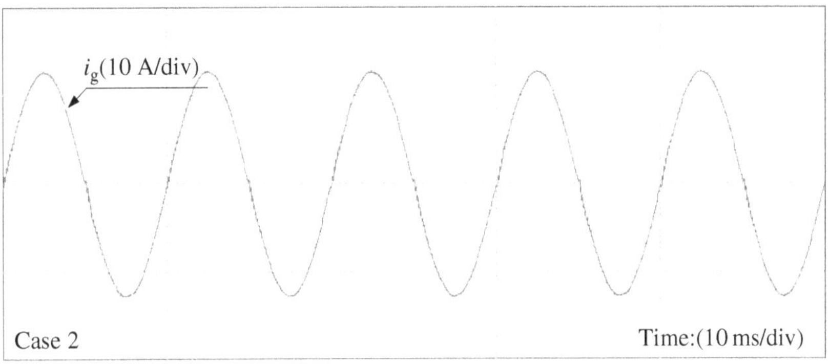

Figure 8.15 Simulated grid-injected current of the *LLCL*-filter-based system with the C_{EMI}, where the parameters of Case 2 are adopted.

respectively, and they become 1 μF and 64 μH, respectively, and NPR_I appears. In this case, f_p was calculated to be 4.477 kHz and NPR_I was 4.477 and 5 kHz.

For Case 3, the intersection frequency becomes 4.7 kHz, which is located in NPR_I. The simulated waveform and FFT waveform of the grid-injected current are shown in Figures 8.16 and 8.17, where the excessive grid-injected current harmonics appear around the frequency of 4.7 kHz.

Based on Case 3, Case 4 adopts the RC damper to eliminate the excessive grid-injected current harmonics, where the simulated grid-injected current in Figure 8.18 shows that the system becomes stable again.

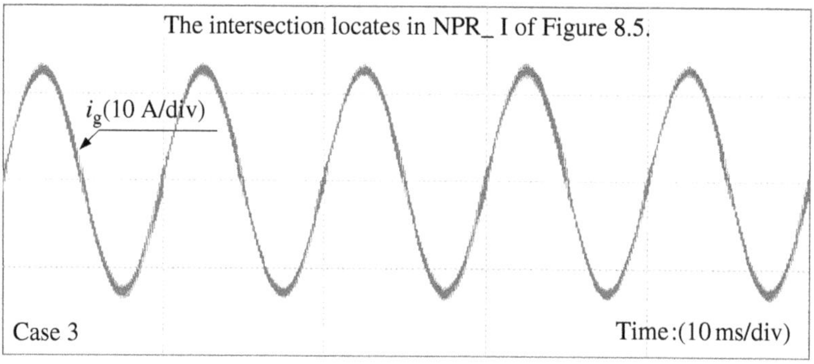

Figure 8.16 Simulated grid-injected current of the *LLCL*-filter-based system with parameters of Case 3, when the intersection is located in NPR_I of Figure 8.5.

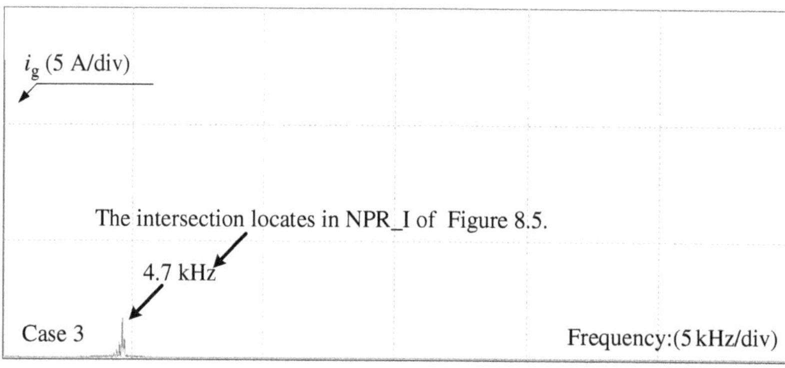

Figure 8.17 Simulated FFT waveform of the grid-injected current with parameters of Case 3, when the intersection is located in NPR_I of Figure 8.5.

The simulation results in Figures 8.13–8.18 agree well with the results of previous theoretical analysis, and the feasibility of the proposed design methods is confirmed.

8.6.2 Experiments

To further verify the theoretical analysis, a 2 kW/220 V prototype based on a DSP (TMS320LF2812) controller was constructed, while the circuit parameters remained the same as those used in the simulations. An external inductor and capacitor were used to emulate the equivalent grid inductance and capacitance,

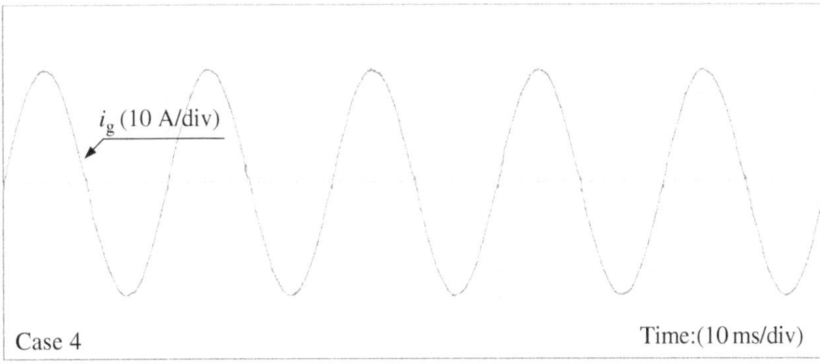

Figure 8.18 Simulated grid-injected current with parameters of Case 4, when the intersection is located in NPR_I of Figure 8.5 and the RC damper is adopted.

respectively. All experimental waveforms were captured using a Yokogawa DL 1640 digital oscilloscope.

8.6.2.1 Experimental Results

In addition, as shown in Table 8.2, Cases 1 and 2 are designed under normal operation, whereas Cases 3 and 4 are designed when the filter parameters drift.

For Case 1, the measured grid-injected current waveform, grid voltage waveform, and FFT waveform of the grid-injected current are shown in Figure 8.19, which match well those of the simulations in Figures 8.13 and 8.14. As expected,

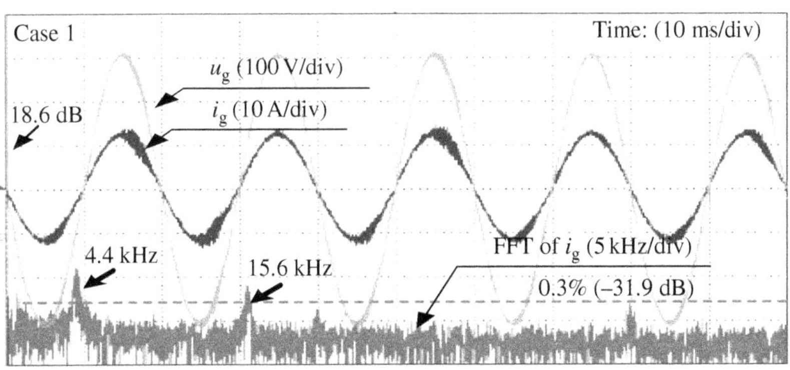

Figure 8.19 Measured grid voltage waveform, grid-injected current waveform, and FFT waveform of the grid-inject current with parameters of Case 1, when the intersection is located in NPR_II of Figure 8.5.

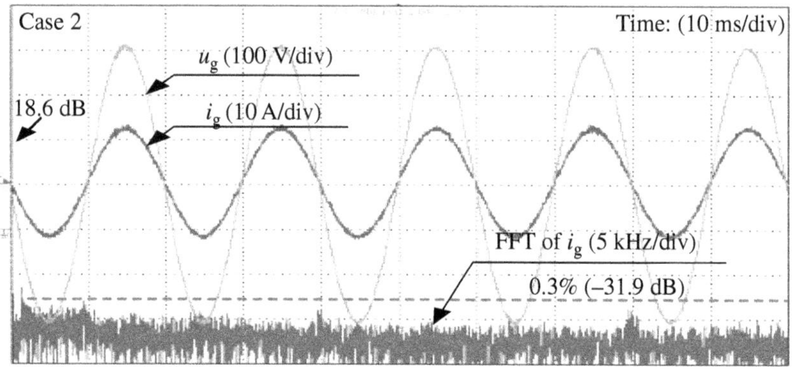

Figure 8.20 Measured grid voltage, grid-injected current, and FFT waveform of the grid-inject current with parameters of Case 2, when the intersection is located in NPR_II of Figure 8.5, and the C_{EMI} = 1 μF is added.

the excessive grid-injected current harmonics appear around 15.6 kHz, which is located in NPR_II.

To avoid the intersection from falling into NPR_II, an extra capacitor C_{EMI} is used as in Case 2. The measured grid-injected current waveform, grid voltage waveform, and FFT waveform of the grid-injected current are shown in Figure 8.20, indicating that the system operates stably again.

However, the filter parameters may drift in actual applications. If C_f and L_f drift +25% and –20%, respectively, NPR_I appears, that is, (4.477, 5) kHz.

For Case 3, the measured grid-injected current waveform, grid voltage waveform, and FFT waveform of the grid-injected current are shown in Figure 8.21,

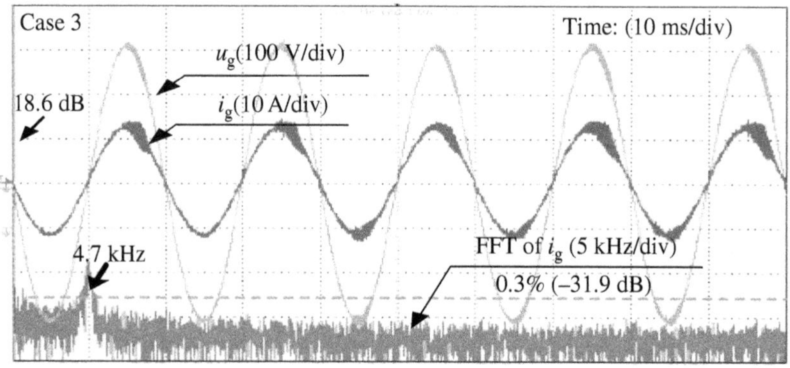

Figure 8.21 Measured grid voltage waveform, grid-injected current waveform, and FFT waveform of the grid-injected current with parameters of Case 3, when the intersection is located in NPR_I of Figure 8.5.

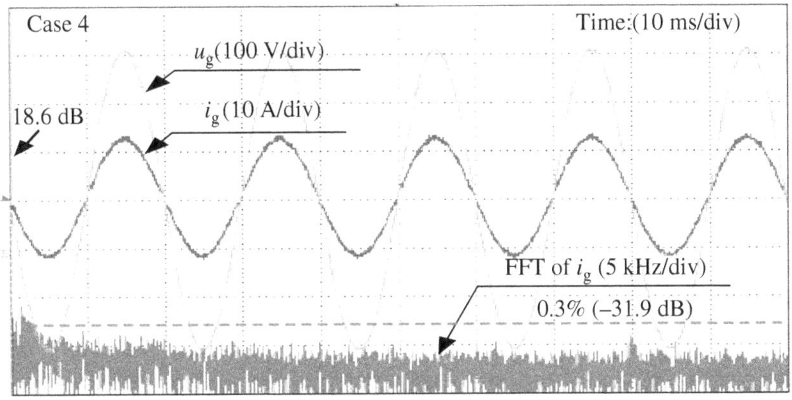

Figure 8.22 Measured grid voltage waveform, grid-injected current waveform, and FFT waveform of the grid-injected current with parameters of Case 4, when the intersection is located in NPR_I of Figure 8.5, and the RC damper is adopted.

which match well those of the simulations in Figures 8.15 and 8.16. The excessive grid-injected current harmonics appear around 4.7 kHz, which is located in NPR_I.

To eliminate the excessive grid-injected current harmonics, the RC damper was adopted as in Case 4. The measured grid-injected current waveform, grid voltage waveform, and FFT waveform of the grid-injected current are shown in Figure 8.22, where the system becomes stable again.

Figure 8.23 shows the dynamic waveform of the grid-injected current of Case 4 when it changes from 12.6 to 6.3 A under the condition that $C_g = 3$ μF and $L_g = 4$ mH. The grid-injected current rapidly follows the grid-injected current reference, which verifies that the system has sufficient control bandwidth under the weakest grid condition.

Figure 8.24 shows the measured current I_d of the R_d–C_d branch of the *LLCL*-filter-based system under the condition that $C_g = 3$ μF and $L_g = 4$ mH. The power loss rate of the RC parallel damper can be calculated according to $(I_d^2 R_d/P_o)\%$, where P_o is the power injected into the grid. The power loss rate was limited to 0.01%, according to the measured current value.

8.6.2.2 Analysis and Discussion

The experimental results show that these are identical to the previous theoretical analysis and simulation results. Moreover, the following still needs to be addressed.

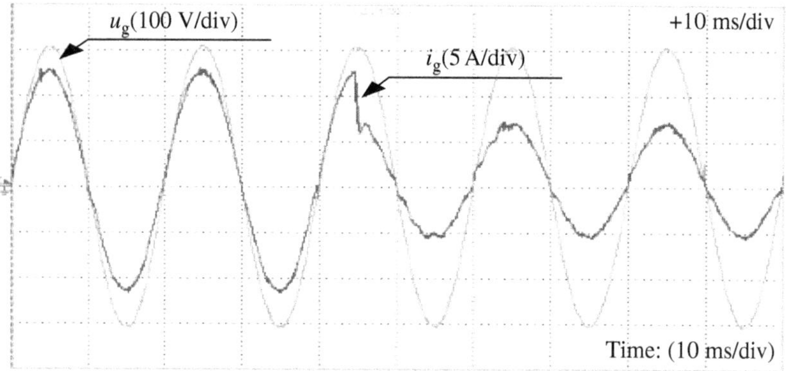

Figure 8.23 Measured grid voltage grid-injected current of Case 4, when the grid-injected current reference is regulated under the condition that $C_g = 3\ \mu F$ and $L_g = 4\ mH$.

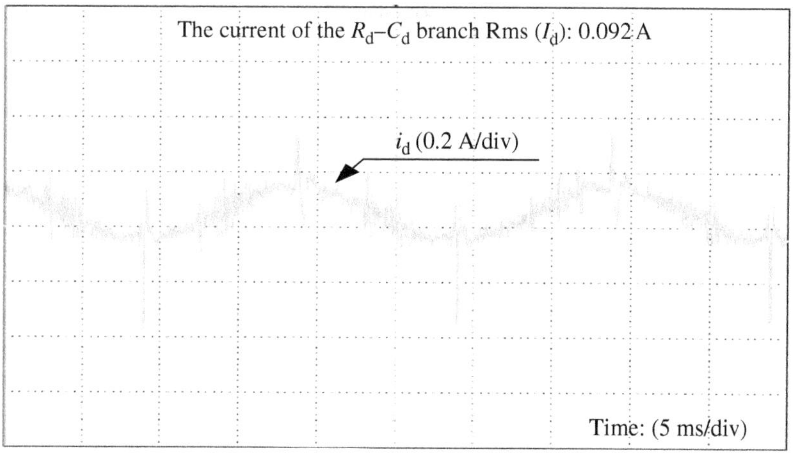

Figure 8.24 Measured current waveform of the C_d–R_d branch of the *LLCL*-filter-based system under the condition that $C_g = 3\ \mu F$ and $L_g = 4\ mH$.

1) Owing to the small values of the damping resistor and capacitor, the extra damping power losses can almost be ignored.
2) Under the tolerance of the industrial power filter, C_f can drift by ±5%, and L_1 and L_f can drift within ±25%. Substituting these tolerances into (8.7), the value of f_p varies between 87.5 and 115.6% of its nominal value f_{p_nom}. The system can still work well with an RC damper, exhibiting the characteristic of high robustness.

3) In fact, there is a trade-off between the control bandwidth and maximum tolerated values of the equivalent C_g and L_g. The larger the multiplication value of $C_g L_g$, the narrower the control bandwidth of the system, where the highest harmonic compensator frequency has to be suppressed correspondingly.

8.7 Summary

This paper first fully analyzes the characteristics of the *LLCL*-filter-based grid-connected inverter output admittance under a wide variation of grid reactance, where some conclusions can be drawn as follows:

1) Within the switching frequency, one or two non-passive regions can be identified depending on the delay time and resonance frequency f_p.
2) When the grid reactance varies over a wide range, the intersection of the inverter output admittance and the grid admittance may fall in the non-passive region.
3) To prevent possible harmonic instability, two basic solutions are recommended: narrowing the non-passive region as much as possible and avoiding the intersection located in the non-passive region.
4) When L_f of the *LLCL* filter is set to zero, the above three points can also be applied to the traditional *LCL* filter-based system.

Second, based on the analysis, the exact and robust parameter design of the *LLCL* filter, the EMI capacitor, and the RC passive damper were introduced in detail. The entire system can operate stably under a wide variation in the grid reactance.

Finally, simulations and experiments were conducted on a 2 kW/220 V *LLCL*-filter-based single-phase grid-connected inverter prototype, and the results confirm the theoretical analysis and feasibility of the proposed measures.

References

1 Enslin, J.H.R. and Heskes, P.J.M. (2004). Harmonic interaction between a large number of distributed power inverters and the distribution network. *IEEE Trans. Power Electron.* 19 (6): 1586–1593.

2 Zhang, S., Jiang, S., Lu, X. et al. (2014). Resonance issues and damping techniques for grid-tied inverters with long transmission cable. *IEEE Trans. Power Electron.* 29 (1): 110–120.

3 Huang, M., Wang, X., Loh, P.C., and Blaabjerg, F. (2016). LLCL-filtered grid converter with improved stability and robustness. *IEEE Trans. Power Electron.* 31 (5): 3958–3967.

4 Yoon, C., Bai, H., Beres, R.N. et al. (2016). Harmonic stability assessment for multiparalleled, grid-tied inverters. *IEEE Trans. Sustainable Energy* 7 (4): 1388–1397.

5 Song, Y., Wang, X., and Blaabjerg, F. (2017). Impedance-based high-frequency resonance analysis of DFIG system in weak grids. *IEEE Trans. Power Electron.* 32 (5): 3536–3548.

6 Bai, H., Wang, X., and Blaabjerg, F. (2017). Passivity enhancement in renewable energy source based power plant with paralleled grid-tied VSIs. *IEEE Trans. Ind. Appl.* 53 (4): 3793–3802.

7 Harnefors, L., Wang, X., Yepes, A.G., and Blaabjerg, F. (2016). Passivity-based stability assessment of grid-tied VSCs – an overview. *IEEE J. Emerg. Sel. Topics Power Electron.* 4 (1): 116–125.

8 Zhang, X., Spencer, J.W., and Guerrero, J.M. (2013). Small-signal modeling of digitally controlled grid-tied inverters with LCL filters. *IEEE Trans. Ind. Electron.* 60 (9): 3752–3765.

9 Harnefors, L., Finger, R., Wang, X. et al. (2017). VSC input-admittance modeling and analysis above the Nyquist frequency for passivity-based stability assessment. *IEEE Trans. Ind. Electron.* 64 (8): 6362–6370.

10 Sun, J. (2011). Impedance-based stability criterion for grid-tied inverters. *IEEE Trans. Power Electron.* 26 (11): 3075–3078.

11 Wen, B., Burgos, R., Boroyevich, D. et al. (2017). AC stability analysis and dq frame impedance specifications in power-electronics-based distributed power systems. *IEEE J. Emerg. Sel. Topics Power Electron.* 5 (4): 1455–1465.

12 Freijedo, F.D., Rodriguez-Diaz, E., Golsorkhi, M.S. et al. (2017). A root-locus design methodology derived from the impedance/admittance stability formulation and its application for LCL grid-tied converters in wind turbines. *IEEE Trans. Power Electron.* 32 (10): 8218–8228.

13 Wu, W., Sun, Y., Lin, Z. et al. (2014). A modified LLCL filter with the reduced conducted EMI noise. *IEEE Trans. Power Electron.* 29 (7): 3393–3402.

14 Xu, J., Yang, J., Ye, J. et al. (2014). An LTCL filter for three-phase grid-tied converters. *IEEE Trans. Power Electron.* 29 (8): 4322–4338.

15 Li, F., Zhang, X., Zhu, H. et al. (2015). An LCL-LC filter for grid-tied converter: topology, parameter, and analysis. *IEEE Trans. Power Electron.* 30 (9): 5067–5077.

16 Wu, W., He, Y., and Blaabjerg, F. (2012). An LLCL power filter for single-phase grid-tied inverter. *IEEE Trans. Power Electron.* 27 (2): 782–789.

17 Wu, W., He, Y., Tang, T., and Blaabjerg, F. (2013). A new design method for the passive damped LCL and LLCL filter-based single-phase grid-tied inverter. *IEEE Trans. Ind. Electron.* 60 (10): 4339–4350.

18 Wu, W., Sun, Y., Huang, M. et al. (2014). A robust passive damping method for LLCL filter based grid-tied inverters to minimize the effect of grid harmonic voltages. *IEEE Trans. Power Electron.* 29 (7): 3279–3289.

19 Wang, J., Yan, J.D., Jiang, L., and Zou, J. (2016). Delay-dependent stability of single-loop controlled grid-tied inverters with LCL filters. *IEEE Trans. Power Electron.* 31 (1): 743–757.

20 Jia, L., Ruan, X., Zhao, W. et al. (2018). An adaptive active damper for improving the stability of grid-tied inverters under weak grid. *IEEE Trans. Power Electron.* 33: 9561–9574.

Part V

Active Damping

9

Active Damping of *LLCL*-Filter Resonance Based on *LC*-Trap Voltage or Current Feedback

9.1 Introduction

Distributed power generation systems powered by clean energy sources, such as solar and wind energy, are becoming popular in the industry [1, 2]. These sources are almost always interfaced by grid converters for conditioning the generated energy before channeling it to the grid. A low-pass power filter is also normally inserted between each converter and grid to attenuate undesired harmonics caused by modulation of the converter. The low-pass filter can be an *L*-filter or a more recent *LCL* filter [3, 4]. For even better attenuation around the switching frequency of the converter, the middle *C* of the *LCL* filter can be replaced by a series *LC*-trap tuned at the switching frequency. The created fourth-order filter is called the *LLCL* filter with even smaller total inductance and capacitance [5–10].

However, similar to the *LCL* filter, research on the *LLCL* filter is still required to solve the resonance problem, which can be resolved by active [11–16] or passive damping [17–21]. Compared with passive damping, active damping avoids the need for additional resistors to reduce the overall system efficiency. Active damping using control algorithms instead of resistors has been extensively studied without any additional power loss. It has the characteristic of flexibility, with many control possibilities available for selection. Among them, the multi-loop active damping methods are the most widely discussed.

With delays in most digitally controlled systems, the single resistor equivalence is no longer valid. More stability-related issues will also surface, and it has been found that delays will help stabilize a single grid current control loop without damping if the overall system resonance frequency is higher than a critical value [22]. In contrast, a single converter current control loop without damping is stable only at a low system resonance frequency. However, these findings about un-damped systems are not very helpful when the grid impedance and hence the system resonance

Control and Filter Design of Single-Phase Grid-Connected Converters, First Edition. Weimin Wu, Frede Blaabjerg, Henry Chung, Yuanbin He, and Min Huang.

Published 2023 by John Wiley & Sons, Inc.

frequency change significantly. It is therefore safer or more robust to include a damper, such as the popularly investigated proportional capacitor current damper for an *LCL* filter. However, the proportional capacitor current damper has been proven to be suboptimal when delays are considered. Instead, the capacitor current should pass through a high-pass filter, which verifies its performance. However, with a high-pass filter, the concern of noise complications is always present.

The same concern may burden an *LLCL* filter when its capacitor current, which is also its *LC*-trap current, is fed back for damping. Various trap current dampers for the *LLCL* filter are investigated in the third section of this chapter. More importantly, the investigation has been expanded to include trap voltage feedback, which, upon analysis, is found to work with even a simple proportional damper. Hence, it is a dynamically fast and less noise-sensitive damper for the *LLCL* filter. Related expectations can similarly be deduced from circuit equivalences derived for all studied dampers, which are convenient tools for practicing engineers. Finally, the expected performance of the damper in the fifth section was verified in the experiment of the grid converter.

9.2 Control of *LLCL*-Filtered Grid Converter

9.2.1 Description and General Control

Figure 9.1 shows a grid-connected converter with an *LLCL* filter comprising converter-side inductance L_1, grid-side inductance L_2, trap inductance L_f, and capacitance C_f. Although the middle *LC*-trap of the filter in the figure is star- or

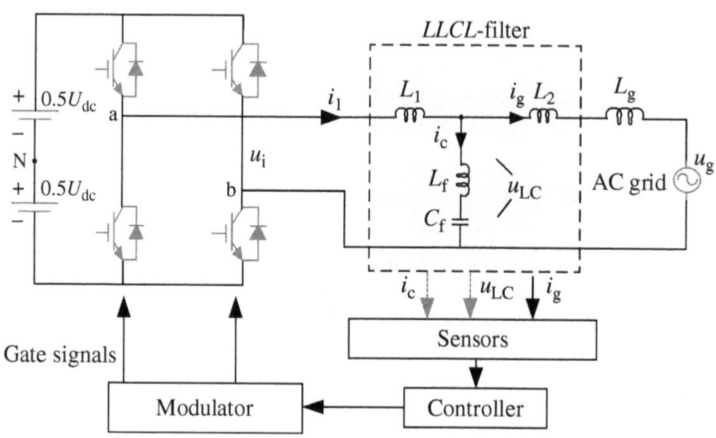

Figure 9.1 Grid-connected converter with an *LLCL*-filter.

wye-configured, it can also be delta-configured, if preferred. However, the star configuration may be more attractive because of its smaller inductance requirement and hence smaller accompanying parasitic resistance. However, the configuration eventually chosen will not affect the active damping conclusions uncovered by the presented analyses. Returning to Figure 9.1, the DC-link voltage of the converter is further assumed to be constant, which, when modulated, produces the immediate phase output voltage u_i and current i_1 of the converter. Other parameters of interest are the trap voltage u_{LC}, trap or capacitor current i_c, grid voltage u_g, grid current i_g, and grid inductance L_g. The values chosen for these parameters, and the other operating parameters of the converter are listed in Table 9.1.

The assembled converter was controlled by a single loop to regulate its grid current. The controller used was a standard proportional-resonant controller with selective harmonic compensation [23]. The transfer function of the controller is provided in (9.1):

$$G_c(s) = K_p + \sum_{h=1,5,7} \frac{K_{ih} s}{s^2 + (\omega_0 h)^2} \tag{9.1}$$

where $\omega_0 = 2\pi f_0$ is the fundamental angular frequency of the grid, K_p is the proportional gain, and K_{ih} is the resonant gain of harmonic order h. For (9.1) to be effective, the grid frequency must be known. To avoid this requirement, a proportional-integral feedforward controller with grid voltage can be used as an alternative. However, notably, both the resonant part of (9.1) and the feeding forward of the grid voltage are employed to improve steady-state tracking [24]. They will not differ greatly in resonance damping, if the same proportional gain is used.

Table 9.1 System parameters.

Symbol	Meaning	Value
U_{dc}	DC-link voltage	175 V
U_g	Grid phase voltage	110 V
L_1	Converter-side	1.8 mH
L_f, C_f	Trap circuit	64 μH, 4 μF
f_{sw}	Switching frequency	10 kHz
f_s	Sampling frequency	10 kHz
f_r	Resonance frequency	2.45 kHz
L_2	Grid-side	2 mH
L_g	Grid inductance (unless stated otherwise)	0 mH
T_s	Sampling period	100 μs

The grid converter in Figure 9.1 is also commonly damped by feeding back a variable [16, 25], which, for the *LCL* filter, is popularly chosen as the capacitor current i_c. The same i_c can be fed back for the *LLCL* filter, but it will be demonstrated later that the trap voltage u_{LC} is a better feedback variable that can preserve both simplicity and robustness. The common purpose is still to damp the resonance at $\omega_r = 2\pi f_r$, whose expression can be extracted from any of the following open-loop transfer functions i_g/u_i, u_{LC}/u_i, and i_c/u_i.

$$G_{u_i \to i_g}(s) = \frac{L_f C_f s^2 + 1}{C_f[L_1 L_2 + (L_1 + L_2)L_f]s(s^2 + \omega_r^2)} \tag{9.2}$$

$$G_{u_i \to u_{LC}}(s) = \frac{L_2 s(L_f C_f s^2 + 1)}{C_f[L_1 L_2 + (L_1 + L_2)L_f]s(s^2 + \omega_r^2)} \tag{9.3}$$

$$G_{u_i \to i_c}(s) = \frac{L_2 s^2}{[L_1 L_2 + (L_1 + L_2)L_f]s(s^2 + \omega_r^2)} \tag{9.4}$$

$$\omega_r = \frac{1}{\sqrt{(L_1 + L_2/L_1 + L_2 + L_f)C_f}} \tag{9.5}$$

Note that (9.2)–(9.5) have considered a zero grid inductance, which, when nonzero, will change L_2 to $L_2 + L_g$. The resonance frequency f_r of the system is then affected by the grid variations.

9.2.2 Block Diagrams of Different Active Dampers

Figure 9.2 shows the control block diagram of the grid converter when its active damping is implemented by feeding back the trap voltage u_{LC}, which can also be fed to a phase-locked loop for synchronizing the desired grid current command i_g^* in the figure with u_{LC}. Thus, an extra sensor is saved. In the meantime, the feedback path with the $1/Z_{eq}(s)$ block should also be ignored first because it will only be explained in the next section. The feedback variable u_{LC} is then passed through a

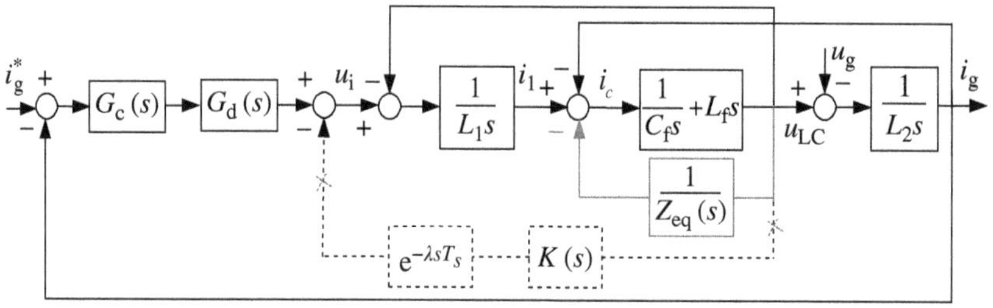

Figure 9.2 Grid current control with a damper based on the trap voltage feedback.

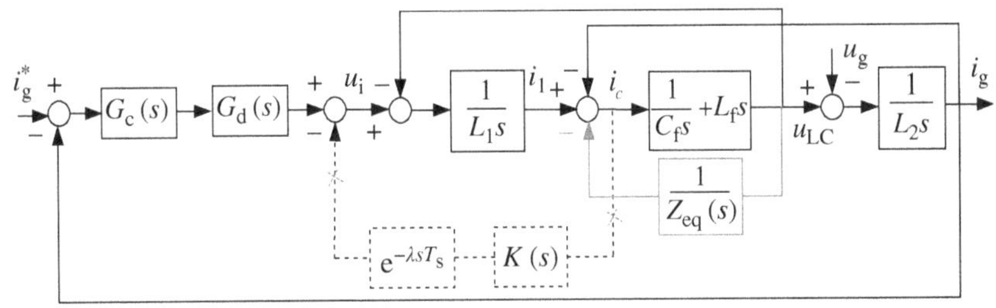

Figure 9.3 Grid current control with a damper based on the trap current feedback.

damper $K(s)$ and a delay block $G_d(s) = e^{-\lambda s T_s}$, which is also added after the grid current controller $G_c(s)$. The delay here is λ, normalized with respect to sampling period T_s. The sampling frequency fs is then given by $1/T_s$, which can be set equal to the converter switching frequency ($\Rightarrow f_s = f_{sw} = 10$ kHz) or higher. Conceptually, a higher sampling frequency gives rise to a smaller and hence less impactful delay. Cases with higher sampling frequencies were thus grouped with analog systems ($\lambda = 0$) in this study [26]. Instead of the trap voltage, the trap current i_c can also be sensed and fed back for resonance damping, as shown in Figure 9.3. Again, the feedback path with the $1/Z_{eq}(s)$ block should be ignored first as it will only be explained in the next section.

9.2.3 Effects of Delay $G_d(s)$

Figure 9.4 shows the Bode plots of the open-loop transfer function $i_g(s)/u_i(s)$ with and without damping. The parameters used for the plots are given in Table 9.1, while the delay of the system was set to $\lambda T_s = 1.5T_s$ or more, precisely T_s plus $0.5T_s$. The first delay of T_s is contributed by the usual process of sampling the data, performing the computation, and updating the signal for pulse-width modulation one sampling period later [27]. After the signal is updated, it is held constant by the digital pulse-width modulator over the next full sampling period. This holding process can be approximated by a zero-order hold with the following transfer functions expressed as both s and $j\omega$ [10, 27]:

$$H_0(s) = \frac{1 - e^{-T_d s}}{s} \tag{9.6}$$

$$H_0(j\omega) = \frac{1 - e^{-j\omega T_s}}{j\omega} = \frac{\sin(0.5\omega T_s)}{0.5\omega} e^{-j0.5\omega T_s} \approx T_s e^{-j0.5\omega T_s} \tag{9.7}$$

Clearly, the $H_0(j\omega)$ function in (9.3) indicates a second delay of $0.5T_s$, which when added to the computation delay of T_s, gives a total delay of $1.5T_s$ or $\lambda = 1.5$. According to Figure 9.4, this delay will not destabilize the system even with

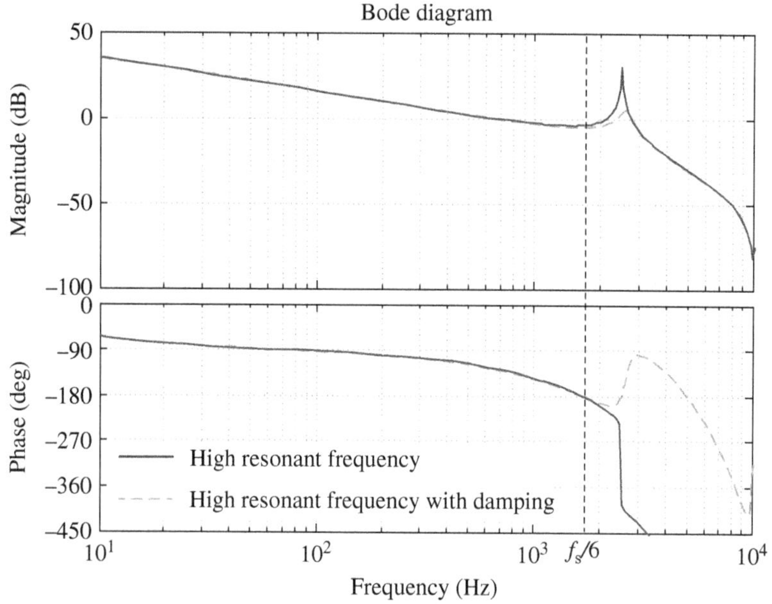

Figure 9.4 Bode plots of open-loop $i_g(s)/u_i(s)$ with and without damping at a high resonance frequency.

no damping added (solid traces in Figure 9.4). The only requirement is to design resonance f_r to be higher than the critical frequency of $f_s/6$ [11], above which, the system phase falls well below −180° caused by the delay. The sudden phase transition at f_r after $f_s/6$ has no prominent influence on the system stability. If proportional trap current damping is added to the system with its resonance still kept above $f_s/6$, the dashed traces in Figure 9.4 show that the system phase crosses −180° twice. The cause is related to the non-minimum-phase (NMP) response, which, in simple terms, denotes an initial decrease before increasing to track the raised command reference. The stability of the system must then be determined by the Nyquist stability criterion, which [10] has proven that for a stable *LCL*-filtered system, its magnitude margins at $f_s/6$ and f_r must be larger and smaller than zero, respectively.

Comparatively, it is advisable for the converter to operate undamped if $f_r > f_s/6$. However, such a condition may be breached if the filter inductances (L_x, $x = 1$, 2, or f), filter capacitance C_f, and grid inductance L_g increase, according to (9.2), but with L_2 replaced by $L_2 + L_g$. Frequency f_r will then move leftward and toward the condition of $f_r < f_s/6$. When this occurs, damping may be unavoidable. Therefore, it might be more attractive to design a maximum $f_r < f_s/6$ at the beginning with damping added, $L_g = 0$ is assumed, and the smallest filter inductances and capacitance are considered (L_x−ΔL_x and C_f−ΔC_f, where Δ represents the percentage variation). As L_x, C_f, and L_g subsequently increase, f_r will only move leftward to the maximum value without breaching the desired condition of $f_r<f_s/6$. Therefore, uniformity and robustness were ensured. In the case of adding more *LC* traps to the *LLCL* filter to create a wider attenuating notch around the converter switching frequency, the principle of ensuring maximum $f_r<f_s/6$ should similarly be applied even with more resonance peaks present.

The level of robustness can be increased by inserting a large shunt capacitance between L_2 and L_g in Figure 9.1. Variations in L_g will not significantly affect the resonance frequency f_r, but at the cost of a shunt capacitance. Alternatively, the critical frequency of $f_s/6$ can be increased, and one way to achieve it is documented in [10], but only for an *LCL* filter with a proportional active damper based on capacitor current feedback. The new critical frequency achieved by [10] is $f_s/4$, which is theoretically obtained by eliminating the time difference between the capacitor current sampling and modulation reference updating instants. Therefore, it requires a unique sampling pattern, which must first be supported by a control platform. To avoid modifying the standard sampling pattern, the next section analyzes a few alternatives whose common basic principle is to increase the critical frequency by reshaping the virtual filter impedance. Such reshaping can be achieved by choosing an appropriate damping function for $K(s)$ in Figure 9.2 or Figure 9.3. Therefore, it is much simpler, with an even higher level of robustness demonstrated (critical frequency higher than $f_s/4$).

9.3 Circuit Equivalences of *LLCL* Active Dampers

9.3.1 General Virtual Impedance Model

To derive the circuit equivalence of the damper shown in Figure 9.2, the usual feedback path through $K(s)$ has been changed by shifting its output node from after $G_d(s)$ to after $1/sL_1$. The modified path through $1/Z_{eq}(s)$ is drawn using the solid line in the figure. A similar shift can also be applied to Figure 9.3, which, in addition to its output node, has its input node shifted from i_c to u_{LC}. The modified damping path through $1/Z_{eq}(s)$ is again drawn using the solid line in the figure. Undoubtedly, the $Z_{eq}(s)$ values for both figures are different; however, in terms of circuit equivalence, they can both be represented by Figure 9.5, where $Z_{eq}(s)$ has been

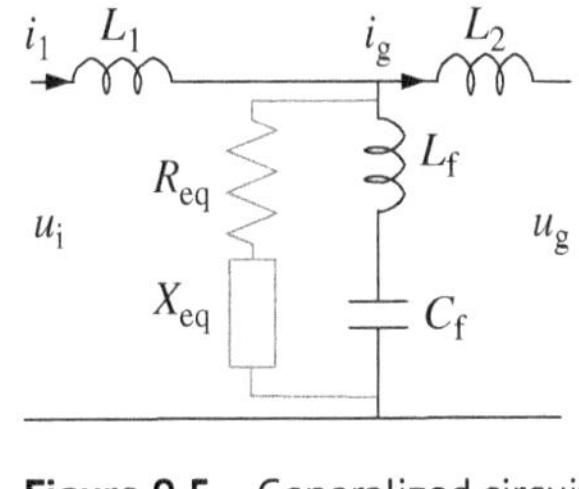

Figure 9.5 Generalized circuit equivalence of active damper.

separated into $R_{eq}(s)$ and $X_{eq}(s)$ in series. The precise forms assumed by $R_{eq}(s)$ and $X_{eq}(s)$ depend on the damping function $K(s)$ and the feedback variable chosen, as explained in the following subsections.

9.3.2 LC-Trap Voltage Feedback

With its *LC*-trap voltage fed back for damping, as shown in Figure 9.2, the generalized expression derived for representing $Z_{eq}(s)$ can be expressed as (9.8):

$$Z_{eq}(s) = s \cdot L_1/K(s) \cdot e^{\lambda s T_s} \tag{9.8}$$

The exponential delay function in (9.4) can be expanded as in (9.9), allowing $Z_{eq}(s)$ to eventually be written as a real and an imaginary term, as in (9.10).

$$e^{j\omega\lambda T_s} = \cos \omega\lambda T + j \sin \omega\lambda T, s \rightarrow j\omega \tag{9.9}$$

$$Z_{eq}(j\omega) = R_{eq}(\omega) + jX_{eq}(\omega) \tag{9.10}$$

The real term is for resistive damping, whereas the imaginary term provides a degree of freedom for shifting the system resonance frequency. The polarity and frequency dependencies of both terms depend on the precise damping function $K(s)$ chosen, which for preserving the simplicity of the overall system, has been limited to the five common functions of k, ks, k/s, $ks/(s+\tau)$, and $k/(s+\tau)$. Parameters k and τ used with these functions represent their damping gain and cut-off angular frequency, respectively. Substituting the functions into (9.8) to (9.10) leads to the R_{eq} and X_{eq} expressions shown in Table 9.2. Table 9.2 can be simplified by setting $\lambda = 0$ to represent an analog system without delay (or a digital system with extremely fast sampling). The simplified R_{eq} and X_{eq} can then be drawn as in Figure 9.6 with R_d and L_d given in (9.11). These representations show that for all dampers, the polarities of R_{eq} and X_{eq} remain unchanged upon fixing k in (9.11). In addition, with a simple k damper, resistive damping was not introduced. Therefore, a simple k damper is not recommended for analog systems.

$$R_d = \frac{L_1}{k}, L_d = \frac{j\omega L_1}{k} \tag{9.11}$$

The scenario is different when a delay is considered, which for a regularly sampled system, is typically given by $\lambda = 1.5$. To illustrate this, a simple k damper is again considered. Its R_{eq} in Table 9.2 is no longer zero, unlike in Figure 9.6a. It can either be positive or negative depending on k, and the angular frequency ω that needs damping, which, for the system shown in Figure 9.1, is its resonance frequency f_r. Preferably, R_{eq} should not be negative because it introduces open-loop right-half-plane (RHP) poles to the grid current control scheme shown in Figure 9.2. The closed-loop response of the scheme will then have NMP characteristics. Therefore, it is recommended to have a positive R_{eq}, which, for the k damper,

Table 9.2 Equivalent Req and X_{eq} of the different dampers based on the trap voltage feedback.

$K(s)$	k	Ks	k/s	$ks/(s+\tau)$	$k/(s+\tau)$
$R_{eq}(\omega)$	$-\frac{L_1\omega}{k}\cdot\sin\lambda\omega T_s$	$\frac{L_1}{k}\cdot\cos\lambda\omega T_s$	$-\frac{L_1\omega^2}{k}\cdot\cos\lambda\omega T_s$	$\frac{L_1\tau}{k}\cdot\cos\lambda\omega T_s-\frac{L_1\omega}{k}\cdot\sin\lambda\omega T_s$	$-\frac{L_1\omega^2}{k}\cdot\cos\lambda\omega T_s-\frac{L_1\tau\omega}{k}\cdot\sin\lambda\omega T_s$
$X_{eq}(\omega)$	$\frac{L_1\omega}{k}\cdot\cos\lambda\omega T_s$	$\frac{L_1}{k}\cdot\sin\lambda\omega T_s$	$-\frac{L_1\omega^2}{k}\cdot\sin\lambda\omega T_s$	$\frac{L_1\tau}{k}\cdot\sin\lambda\omega T_s+\frac{L_1\omega}{k}\cdot\cos\lambda\omega \mathrm{T}_s$	$-\frac{L_1\omega^2}{k}\cdot\sin\lambda\omega \mathrm{T}_s+\frac{L_1\tau\omega}{k}\cdot\cos\lambda\omega T$
f_n	$f_s/3$	$f_s/6$	$f_s/6$	see Figure 9.7a	see Figure 9.7b

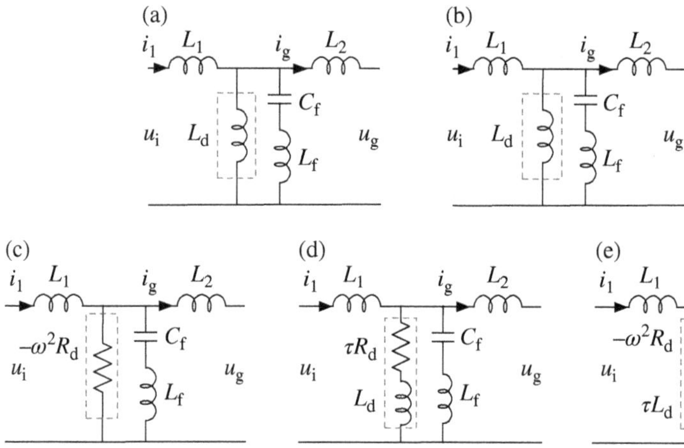

Figure 9.6 Circuit equivalences of (a) k; (b) ks; (c) k/s; (d) $ks/(s + \tau)$; and (e) $k/(s + \tau)$ dampers based on trap voltage feedback and no delay.

can be achieved by setting either of the conditions in (9.12). However, the condition with $f_r<f_s/3$ is preferred because of the uniformity reasons explained in Section 9.2.3.

$$k < 0 \text{ and } \omega = 2\pi f_r < 2\pi f_s/3, \text{or } k > 0 \text{ and } \omega = 2\pi f_r > 2\pi f_s/3 \tag{9.12}$$

Thus, the critical frequency is $f_n = f_s/3$ for the simple k-damper. The same analysis can be performed using the other four dampers. Their respective critical frequencies f_n are summarized in Table 9.2, where it should be noted that the high-pass $ks/(s + \tau)$ and low-pass $k/(s + \tau)$ dampers do not have a fixed f_n. Instead, their f_n increases with their angular cut-off frequencies $\tau = 2\pi f_{nc}$, as shown in Figure 9.7.

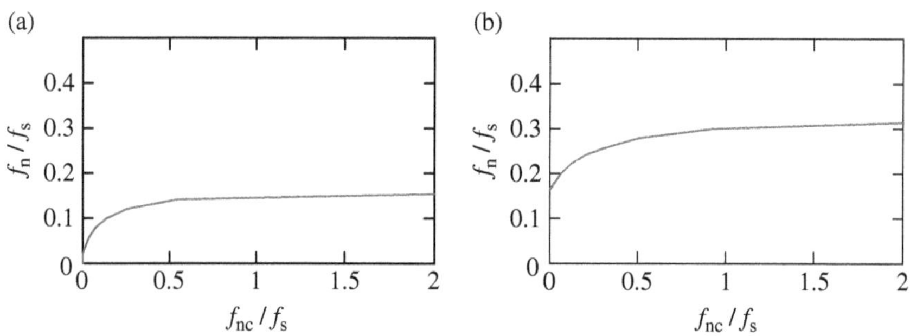

Figure 9.7 Critical frequency f_n versus cutoff frequency $f_{nc} = \tau/(2\pi)$ of the (a) $ks/(s + \tau)$ and (b) $k/(s + \tau)$ dampers based on the trap voltage feedback.

Regardless, f_n of the $k/(s+\tau)$ damper is always higher than that of the $ks/(s+\tau)$ damper when using the same f_{nc}. The former is thus preferred as its f_n approaching that of the simple k damper as f_{nc} increases, as shown in Figure 9.7b. Both k and $k/(s+\tau)$ dampers are therefore more robust if the trap voltage of the regularly sampled system is fed back. This is despite the fact that the k damper was proven earlier to be unsuitable for an analog system (see Figure 9.6a).

To further illustrate the removal of negative resistance and hence NMP response, open-loop Bode plots of Figure 9.2 are plotted with $K(s) = k$ considered as an example. The obtained plots are shown in Figure 9.8 for three different values of k. With $k < 0$, the two representative phase plots clearly decrease monotonically with their phase values at f_r well below −180°. Therefore, they are similar to the solid traces drawn in Figure 9.4, with no open-loop RHP poles and hence no NMP influences. However, notably, out of the two negative k values, only the case with $k = -0.3$ is stable because of its small positive phase margin. The other with $k = -1$ is unstable because its phase has fallen below −180° at the gain crossover frequency. However, such instability is not foretold by (9.12) and its accompanying circuit equivalence because they do not consider the grid current controller $G_c(s)$. By itself, (9.12) only helps in deciding the polarity of k and the placement of f_r to avoid the NMP response. How high the magnitude of k can be, while avoiding instability, is determined by the root-locus analysis presented in Section 9.4.

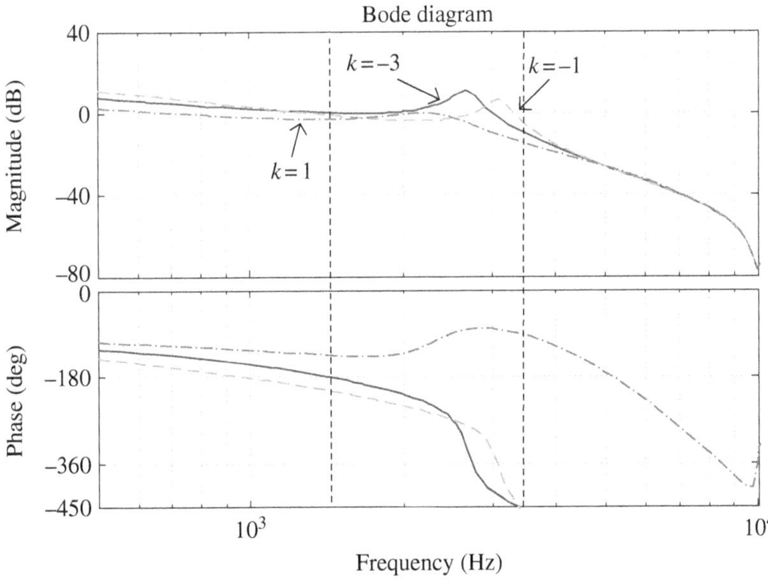

Figure 9.8 Bode plots of Figure 9.2 with the k damper.

9.3.3 LC-Trap Current Feedback

Clearly, the k damper provides the fastest and most robust performance when used with a regularly sampled system with LC-trap voltage feedback. The comparison with the popular capacitor or trap current damper is now investigated, beginning with the generalized expression in (9.13) for representing $Z_{eq}(s)$ in Figure 9.3.

$$Z_{eq} = \frac{L_1(1 + s^2 L_f C_f)}{C_f K(s)} \cdot e^{\lambda s T_d} \tag{9.13}$$

Substituting (9.9) into (9.13) leads to the expressions in Table 9.3 and (9.14) for computing R_{eq} and X_{eq} related to the trap current feedback.

$$A = \frac{L_1(1 - \omega^2 L_f C_f)}{C_f k}, \quad B = \frac{L_1(1 - \omega^2 L_f C_f)}{\omega C_f k}, \quad C = \frac{L_1 \omega}{C_f k} - \frac{L_1 L_f \omega^3}{k} \tag{9.14}$$

It can be observed from (9.14) that A, B, and C are always greater than zero before the converter switching frequency ($\omega < 2\pi f_{sw} = 2\pi f_s = 1/\sqrt{L_f C_f}$). Therefore, they are treated as positive because the usual controllable range is only up to the Nyquist frequency, which is half the sampling frequency. The same expressions in (9.14) can similarly be used with an LCL filter with capacitor current feedback, but only after setting L_f to zero. With a non-zero L_f, the performances of both filters are rightfully different, even though they can be close when L_f is small. However, this depends on the converter switching frequency, from which L_f is computed. With delay then set to $\lambda = 0$ for representing an analog system, Figure 9.9 shows the circuit equivalences obtained with the five damping functions of k, ks, k/s, $ks/(s + \tau)$, and $k/(s + \tau)$. The expressions for R_v, L_v, L_{v1}, and C_v in the figure are given in (9.15), which are clearly positive when k is positive and $0<\omega<2\pi f_s$.

$$R_v = \frac{L_1(1 - \omega^2 L_f C_f)}{C_f k}, \; L_v = j\left(\frac{L_{1\omega}}{C_f k} - \frac{L_1 L_f \omega^3}{k}\right), \; L_{v1} = \frac{j\omega L_1 L_f}{k}, \; C_v = \frac{L_1}{j\omega C_f k} \tag{9.15}$$

Table 9.3 Equivalent R_{eq} and X_{eq} of different dampers based on the trap current feedback.

K(s)	*k*	*Ks*	*k/s*	*ks/(s + τ)*	*k/(s + τ)*
$R_{eq}(\omega)$	$A \cos \lambda\omega T$	$B \sin \lambda\omega T$	$-C \sin \lambda\omega T$	$A \cos \lambda\omega T + B\tau \sin \lambda\omega T$	$\tau A \cos \lambda\omega T - C \sin \lambda\omega T$
$X_{eq}(\omega)$	$A \sin \lambda\omega T$	$-B \cos \lambda\omega T$	$C \cos \lambda\omega T$	$A \sin \lambda\omega T - \tau B \cos \lambda\omega T$	$\tau A \sin \lambda\omega T + C \cos \lambda\omega T$
f_n	$f_s/6$	$f_s/3$	$f_s/3$	see Figure 9.10a	see Figure 9.10b

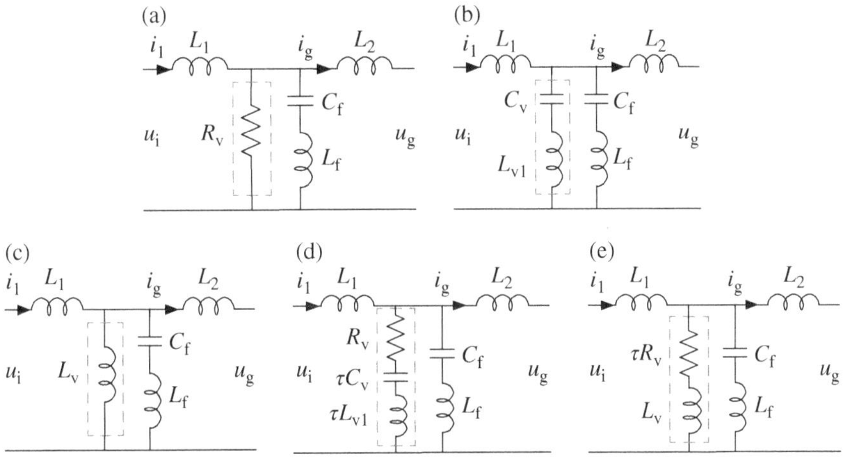

Figure 9.9 Circuit equivalences of (a) k; (b) ks; (c) k/s; (d) $ks/(s + \tau)$; and (e) $k/(s + \tau)$ dampers based on the trap current feedback and no delay.

The resistance in Figure 9.9a is also positive, implying that the k damper can be used with an analog system, where the trap current is fed back.

With a delay of $\lambda = 1.5$ now considered, Figure 9.9 is no longer applicable because R_{eq} can be either positive or negative depending on the chosen parameters. To illustrate this, a simple k damper is again considered. Its R_{eq} is positive only when either condition in (9.16) is satisfied.

$$k > 0 \text{ and } \omega = 2\pi f_r < 2\pi f_s/6, \text{ or } k < 0 \text{ and } \omega = 2\pi f_r > 2\pi f_s/6 \tag{9.16}$$

The critical frequency is thus $f_n = f_s/6$, which is visibly smaller than that of the LC-trap voltage feedback through the same k damper. To increase it to $f_s/3$, ks ($k > 0$) and k/s ($k < 0$) dampers should be used, which when summed in proportion to the L_f/C_f ratio and multiplied with the trap current, give the trap voltage discussed in Subsection 9.3.2. Practically, ks and k/s dampers are, however, not encouraged because the former can cause noise amplification, whereas the latter depends on the initial point of integration. Therefore, they are replaced by high-pass $ks/(s + \tau)$ and low-pass $k/(s + \tau)$ dampers, whose critical frequencies f_n are no longer fixed. Instead, they vary with the cutoff frequencies $\tau = 2\pi f_{nc}$, as shown in Figure 9.10. In particular, the $ks/(s + \tau)$ damper always has a higher critical frequency f_n than fs/6, which eventually saturates at $f_s/3$ as f_{nc} increases. Therefore,

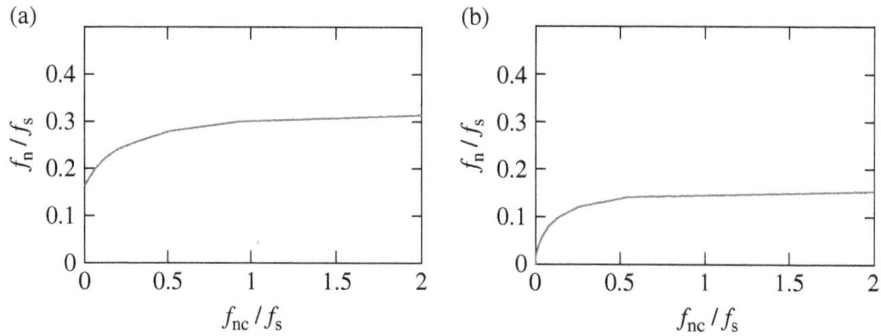

Figure 9.10 Critical frequency f_n versus cutoff frequency $f_{nc} = \tau/(2\pi)$ of the (a) $ks/(s + \tau)$ and (b) $k/(s + \tau)$ dampers based on the trap current feedback.

it is a more robust damper in terms of avoiding the NMP response when compared with the k damper.

9.4 Z-Domain Root-Locus Analysis

Before beginning, it is helpful to clarify that the purpose of Section 9.3 is to derive the circuit equivalences to show how different active dampers modify the original *LLCL* filter. Because the actual filter components are continuous and a discrete power filter does not exist physically, the equivalent circuit derivations are presented in the s-domain with the delay block represented by an exponential term. The overall discrete control design can, however, be performed in the s-domain before discretization or directly in the z-domain. The latter is generally recognized as more precise; hence, it is performed in this section for direct programming to a digital signal processor.

9.4.1 Z-Domain Transfer Functions

To connect the discrete grid current controller $G_c(z)$ and damper $K(z)$ to the continuous *LLCL* and grid parameters, a zero-order-hold block is inserted between them, as shown in Figure 9.11. The resulting discrete open-loop transfer functions derived for representing the trap voltage and trap current dampers are then provided in (9.17) and (9.18), respectively.

$$\frac{i_g(z)}{i_g^*(z)} = G_c(z) \cdot \frac{G_{u_i \to u_{LC}}(z)}{z + K(z) G_{u_i \to u_{LC}}(z)} \cdot \frac{i_g(z)}{u_{LC}(z)} \tag{9.17}$$

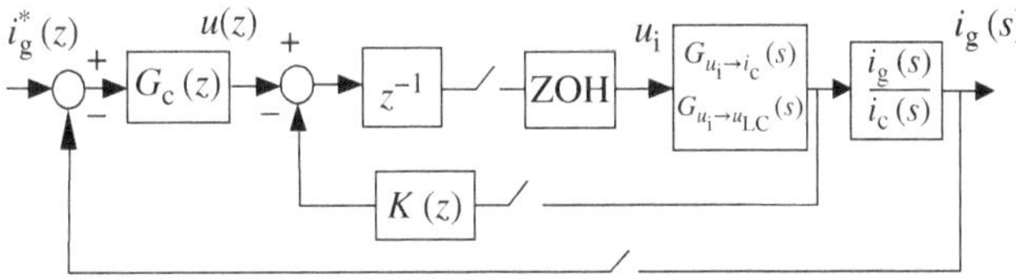

Figure 9.11 Discretized grid current control scheme with active damping.

$$\frac{i_g(z)}{i_g^*(z)} = G_c(z) \cdot \frac{G_{u_i \to i_c}(z)}{z + K(z)G_{u_i \to i_c}(z)} \cdot \frac{i_g(z)}{i_c(z)} \tag{9.18}$$

where $K(z)$ discretized by the Tustin method can be k, $\frac{2\text{x}10^4\text{x}k(z-1)}{z+1}$, $\frac{5\text{x}10^{-5}\text{x}k(z+1)}{z-1}$, $\frac{6.67\text{x}10^{-1}\text{x}k(z-1)}{z-0.3333}$ or $\frac{3.33\text{x}10^{-5}\text{x}k(z+1)}{z-0.3333}$ when $\tau = 10\,000$ and $\lambda = 1.5$. The controller $G_c(z)$ must be discretized according to [28]; however, for the following evaluation, only the simplified form of $G_c(z) = K_p$ is used because the resonant terms in (9.1) will not influence the system response at the resonance frequency f_r.

9.4.2 Root-Locus Analyses with Different Active Dampers

The purpose of this study is to demonstrate how an active damper can stabilize a system without damping. Thus, the initial step is to select a sufficiently high K_p that will marginally push the system in Figure 9.1 to instability when undamped. This means placing the poles on the z-domain unit circle when $k = 0$ to represent no damping. The value was found to be $K_p = 23.9$.

Active damping is then added by increasing $|k|$ with either positive or negative polarity. The former is represented by solid root loci drawn in Figure 9.12, whereas the latter is represented by dashed loci. Although increasing $|k|$ can stabilize the system, the instability will resurface when $|k|$ becomes too high. This common trend can be seen in Figure 9.12, where the root loci of the five practically feasible dampers are shown with either trap voltage or current feedback.

To be more specific, Figure 9.12a shows the k-damped converter with trap-voltage feedback remaining stable only in the range of $-0.3 < k < 0$. As a negative k and with $f_r < f_s/3$ based on the parameters listed in Table 9.1, the converter will also not experience NMP response, as understood from (9.12). The same k damper can equally stabilize the system when the trap current is fed back. However, the stable range is now $0 < k < 11.6$, which unfortunately produces an NMP response because k is positive and $f_r > f_s/6$ (see (9.11)). Unlike the k damper, the other three plots in Figure 9.12 for representing low-pass and high-pass dampers with either trap voltage or current feedback have more loci, introduced by varying their

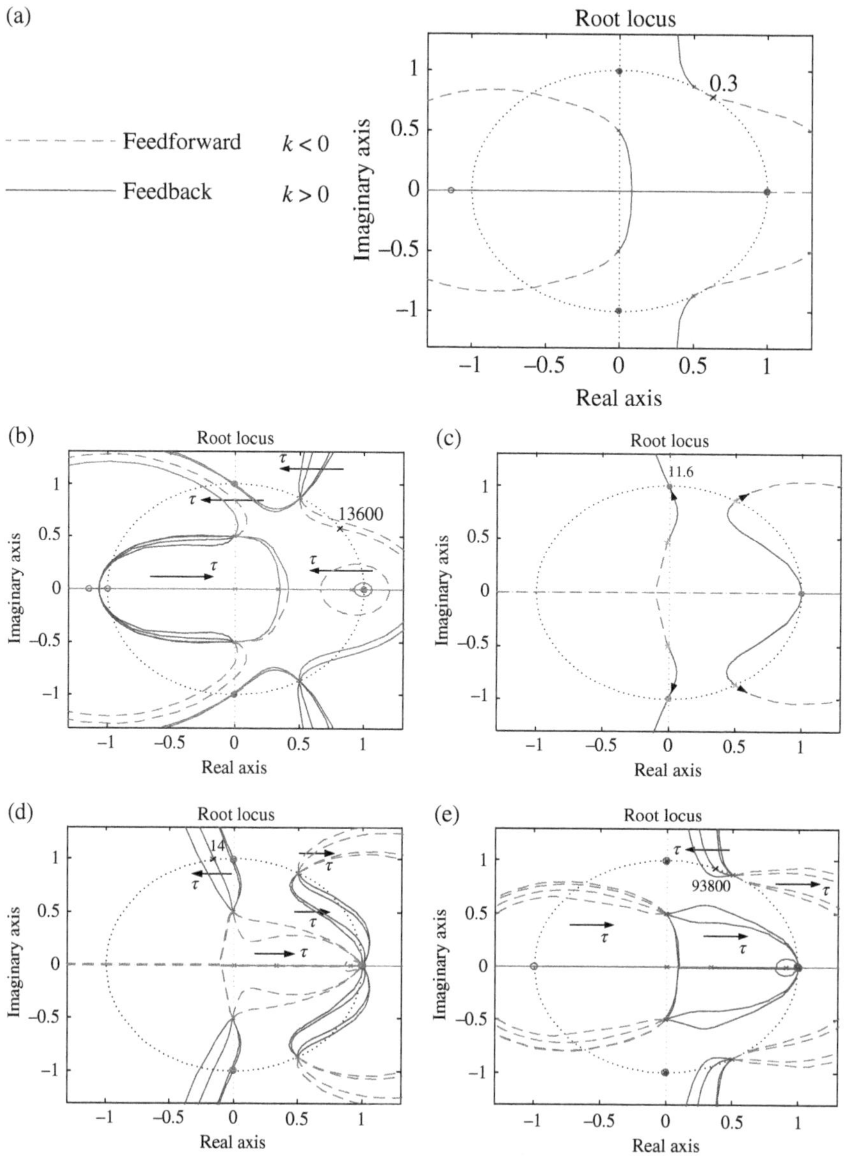

Figure 9.12 Root loci of (a) k damper based on the trap voltage feedback; (b) $k/(s + \tau)$ damper based on the trap voltage feedback; (c) k damper based on the trap current feedback; (d) $ks/(s + \tau)$ based on the trap current feedback; and (e) $k/(s + \tau)$ damper based on the trap current feedback obtained by varying k with $K_p = 23.9$.

Table 9.4 Comparison of different active dampers.

Sensed variable	Trap voltage, u_{LC}		Trap current, i_c		
Damper type	K	Low-pass, $k/(s+\tau)$	k	High-pass, $ks/(s+\tau)$	Low-pass, $k/(s+\tau)$
Stable range	$-0.3 < k < 0$	Negative k Lower limit dependent on τ	$0 < k < 11.6$	Positive k Upper limit dependent on τ	Positive k Upper limit dependent on τ
Critical Frequency, f_n	$f_s/3$	Between $f_s/6$ and $f_s/3$	$f_s/6$	Between $f_s/6$ and $f_s/3$	Between 0 and $f_s/6$
Features	Simple	Tunable with τ	Simple	Tunable with τ, but may have noise complication	Tunable with τ

angular cutoff frequency τ. Each of them has its own stable range for k, even with different polarities. Therefore, they are suitable dampers whose stable k ranges can be widened by increasing τ. However, with the NMP criterion included, Figure 9.10b shows that the low-pass damper with trap current feedback is the least preferred because it has the smallest critical frequency f_n ($<f_s/6$).

9.4.3 Comparison

For easier comparison, results from analyses performed in Sections 9.3 and 9.4 are summarized in Table 9.4. In terms of simplicity and faster dynamics, the k damper with trap voltage feedback is preferred, whereas in terms of flexible tuning (with τ), a low-pass damper with trap voltage feedback is recommended. The high-pass damper with trap current feedback can also be an option, but the presence of a high-pass filter will always lead to the concern of noise complications depending on the operating conditions. Therefore, its use must be slightly more cautious.

9.5 Experimental Verification

The experimental setup consisted of a 1 kW Danfoss FC302 converter connected to the grid through an isolating transformer. The DC link of the converter is tied to a Delta Elektronika power source, and its control is realized with a dSPACE DS1103 controller. The parameters used for the setup were similar to those used for the analyses and are listed in Table 9.1. Similar to the root-locus analyses, each experiment was started with no damping ($k = 0$), $K_p = 23.9$, and $L_g = 0$. Thus, the converter is marginally stable with observable oscillations. After introducing damping ($k \neq 0$), the oscillation should attenuate rapidly if the damper is stable and experiences no NMP response. Otherwise, the oscillation will continue or amplify if the system is unstable, or will increase before gradually diminishing if the system has NMP characteristics.

With this understanding, the experimental trap voltage u_{LC} and grid current i_g waveforms obtained with dampers based on trap voltage feedback are shown in Figure 9.13. In particular, Figure 9.13a shows that with $k = -0.25$ (>-0.3 from Table 9.4), the system is stable. Moreover, because $f_r < f_s/3$ (see Table 9.1), the system does not experience NMP response. Therefore, the initial oscillation rapidly diminishes. The same favorable response is observed in Figure 9.13c, where the results from the low-pass damper are shown. The parameters chosen for the low-pass damper were $k = 12\,200$ (positive) and $\tau = 10\,000$. However, an unstable response is observed with the k damper in Figure 9.13b because of its positive k value intentionally chosen for illustration (see Figure 9.12a and stable k range in Table 9.4).

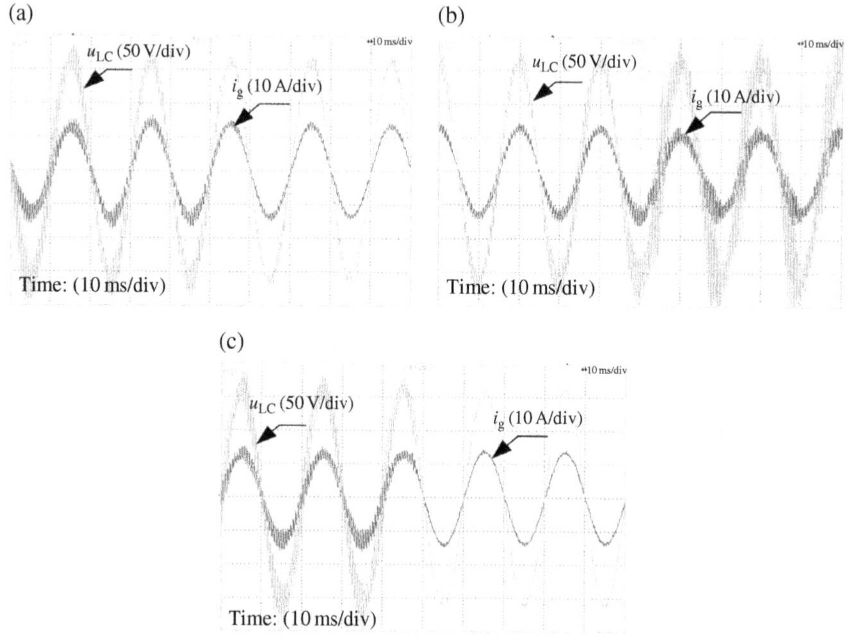

Figure 9.13 Trap voltage u_{LC} and grid current i_g obtained with (a) negative k; (b) positive k; and (c) low-pass $k/(s+\tau)$ dampers (trap voltage feedback and $L_g = 0$).

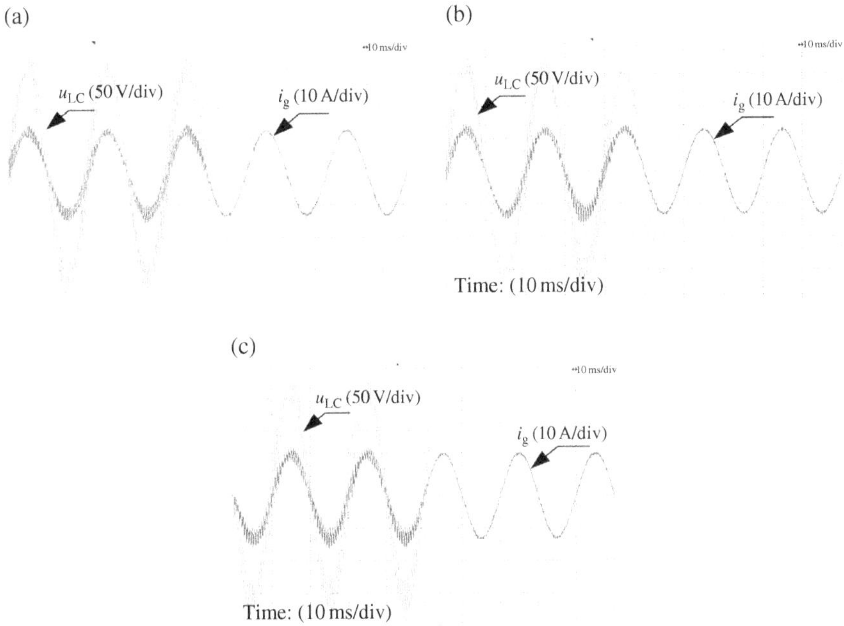

Figure 9.14 Trap voltage u_{LC} and grid current i_g obtained with (a) k; (b) high-pass $ks/(s+\tau)$; and (c) low-pass $k/(s+\tau)$ dampers (trap current feedback and $L_g = 0$).

The same analysis can be performed with Figure 9.14, based on the trap current feedback. All dampers tested were stable, but the k and low-pass dampers were shown to have NMP responses. The reason is related to their low critical frequency, expressed as $f_n = f_s/6$, which, according to values from Table 9.1, is smaller than f_r. Based on Subsection 9.3.3, these dampers are thus NMP, but have slightly different response characteristics owing to their different damping functions. To illustrate, Figure 9.14a, c are compared, where it can be seen that with the dynamically fast k damper, the grid current oscillation in Figure 9.14a surges prominently before diminishing rapidly. However, the surge in the grid current is less obvious with the slower low-pass damper, as shown in Figure 9.14c. The oscillation also diminishes significantly slower.

Figure 9.15 next shows the results from three dampers after increasing the grid inductance L_g to 4.8 mH. The converter-side inductance L_1 has also been increased to 3.6 mH with the other filter parameters remaining unchanged. The purpose is to significantly bring down the resonance frequency f_r to 1.62 kHz according to (9.2), but with L_2 replaced by $L_2 + L_g$. This value is closer to $f_n = f_s/6$ of the k damper with trap current feedback, which according to Table 9.3, will result in zero resistive damping ($R_{eq} \approx 0$). Its results in Figure 9.15b are, therefore, oscillatory. In contrast,

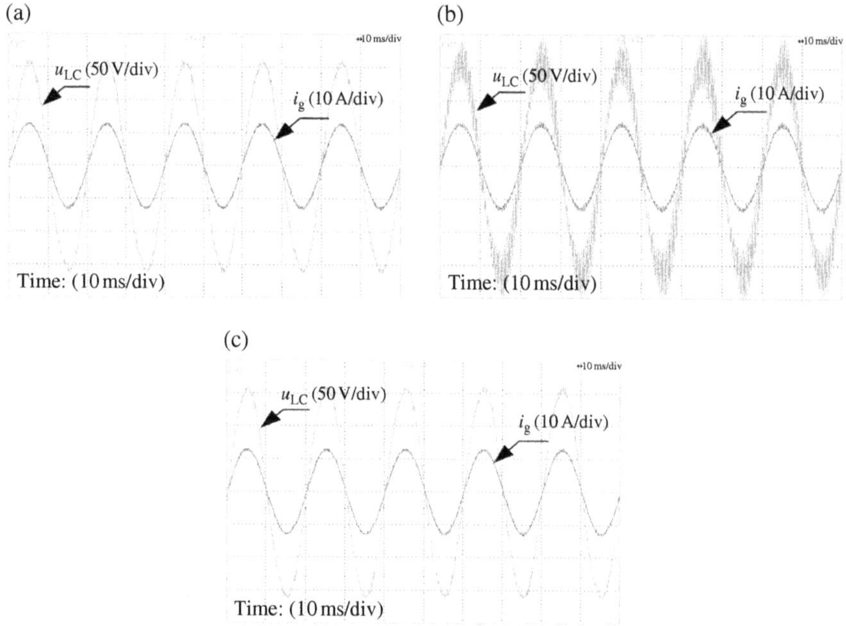

Figure 9.15 Trap voltage u_{LC} and grid current i_g obtained with a (a) k damper and trap voltage feedback; (b) k damper and trap current feedback; and (c) high-pass $ks/(s + \tau)$ damper and trap current feedback (L_g = 4.8 mH).

results in Figure 9.15a for the k damper with trap voltage feedback and Figure 9.15c for the high-pass damper with trap current feedback are not degraded. This common reason is related to their higher critical frequency ($f_n = f_s/3$), which will never be exceeded regardless of how L_g varies. They are therefore robust dampers that can safely be used with a grid converter.

9.6 Summary

This chapter analyses different active dampers based on their non-minimum-phase responses and stability criteria. For the former, different circuit equivalences of the dampers with and without delays were developed. The insights drawn are that an optimal damper designed for an analog system may become suboptimal when used with a regularly sampled system because of delays. A few suitable dampers for the regularly sampled system were identified. They are simple k and low-pass dampers with trap voltage feedback. Both dampers have a high critical frequency and are hence more robust to variations in the grid and filter parameters.

Analyses are then performed to identify their respective stable k ranges, which for the low-pass damper with trap voltage feedback, can be widened by increasing its cutoff frequency. However, this flexibility is at the expense of slower dynamics. The experimental results confirmed these expectations and the unmatched fast dynamics of a simple k damper with trap voltage feedback.

References

1 Blaabjerg, F., Teodorescu, R., Liserre, M., and Timbus, A. (2006). Overview of control and grid synchronization for distributed power generation systems. *IEEE Trans. Ind. Electron.* 53 (5): 1398–1409.

2 Liu, Y., Farnell, C., Zhang, H. et al. (2014). A silicon carbide fault current limiter for distribution systems. *Proceedings of the Energy Conversion Congress and Exposition (ECCE)*, Pittsburgh, PA (14–18 September 2014)

3 Liserre, M., Blaabjerg, F., and Hansen, S. (2005). Design and control of an LCL-filter-based three-phase active rectifiers. *IEEE Trans. Ind. Appl.* 41 (5): 1281–1291.

4 Liserre, M., Teodorescu, R., and Blaabjerg, F. (2006). Stability of photovoltaic and wind turbine grid-connected inverters for a large set of grid impedance values. *IEEE Trans. Power Electron.* 21 (1): 263–272.

5 Jiao, F.L. (2018). LCL filter design and inductor current ripple analysis for 3- level NPC grid interface converter. *IEEE Trans. Power Electron.* 30: 4659–4668.

6 Wu, W., He, Y., and Blaabjerg, F. (2012). An LLCL power filter for single-phase grid-tied inverter. *IEEE Trans. Power Electron.* 27 (2): 782–789.

7 Huang, M., Wu, W., Yang, Y., and Blaabjerg, F. (2013). Step by step design of a high order power filter for three-phase three-wire grid-connected inverter in renewable energy system. *Proceedings of the PEDG*, pp. 1–8, Rogers, AR (8–11 July 2013). New York: IEEE.

8 Cantarellas, A.M., Rakhshani, E., Remon, D., and Rodriguez, P. (2013). Design of the LCL+trap filter for the two-level VSC installed in a large-scale wave power plant. *Proceeding of the ECCE*, pp. 707–712, Denver, CO (15–19 September 2013). New York: IEEE.

9 Dai, K., Duan, K., Wang, X. (2012). Yong Kang application of an LLCL filter on three-phase three-wire shunt active power filter. *Proceedings of the IEEE INTELEC2012*, pp. 1–5, Scottsdale, AZ (September 2012). New York: IEEE.

10 Bloemink, J.M. and Green, T.C. (2011). Reducing passive filter sizes with tuned traps for distribution level power electronics. *Proceedings of the IEEE EPE*, pp. 1–9, Birmingham, UK (August 2011). New York: IEEE.

11 Bao, C., Ruan, X., Wang, X. et al. (2014). Step-by-step controller design for LCL-type grid-connected inverter with capacitor–current-feedback active-damping. *IEEE Trans. Power Electron.* 29 (3): 1239–1253.

12 Pan, D., Ruan, X., Bao, C. et al. (2014). Capacitor-current-feedback active damping with reduced computation delay for improving robustness of LCL-type grid-connected inverter. *IEEE Trans. Power Electron.* 29 (7): 3414–3427.

13 Parker, S., McGrath, B., and Holmes, D.G. (2014). Regions of active damping control for LCL filters. *IEEE Trans. Ind. App.* 50 (1): 424–432.

14 Wang, X., Blaabjerg, F., and Loh, P.C. (2014). Virtual RC damping of LCL-filtered voltage source converters with extended selective harmonic compensation. *IEEE Trans. Power Electron.* 30: 4726–4737.

15 Wang, X., Blaabjerg, F., and Loh, P.C. (2014). Analysis and design of grid-current-feedback active damping for LCL resonance in grid-connected voltage source converters. *Proceedings of the IEEE ECCE*, pp. 373–380, Pittsburgh, PA (14–18 September 2014). New York: IEEE.

16 Dannehl, J., Fuchs, F.W., and Hansen, S. (2010). Investigation of active damping approaches for PI-based current control of grid-connected pulse width modulation converters with LCL filters. *IEEE Trans. Ind. App.* 46 (4): 1509–1517.

17 Peña-Alzola, R., Liserre, M., Blaabjerg, F. et al. (2013). Analysis of the passive damping losses in LCL-filter-based grid converters. *IEEE Trans. Power Electron.* 28 (6): 2642–2646.

18 Channegowda, P. and John, V. (2010). Filter optimization for grid interactive voltage source inverters. *IEEE Trans. Ind. Electron.* 57 (12): 4106–4114.

19 Wu, W., He, Y., and Blaabjerg, F. (2013). A new design method for the passive damped LCL- and LLCL-filter based single-phase grid-tied inverter. *IEEE Trans. Ind. Electron.* 60 (10): 4339–4350.

20 Wu, W., Huang, M., Sun, Y. et al. (2012). A composite passive damping method of the LLCL-filter based grid-tied inverter. *Proceedings of the PEDG*, pp. 759–766, Aalborg, Denmark (June 2012). New York: IEEE.

21 Huang, M., Loh, P.C., Wu, W. and Blaabjerg, F. (2014). Stability analysis and active damping for LLCL-filter based grid-connected inverters. *Proceedings of the IPEC*, pp. 2610–2617, Hiroshima, Japan (18–21 May 2014). New York: IEEE.

22 Zou, C., Liu, B., Duan, S., and Li, R. (2014). Influence of delay on system stability and delay optimization of grid-connected inverters with LCL filter. *IEEE Trans. Ind. Info.* 10 (3): 1775–1784.

23 Jia, Y., Zhao, J., and Fu, X. (2014). Direct grid current control of LCL-filtered grid-connected inverter mitigating grid voltage disturbance. *IEEE Trans. Power Electron.* 29 (3): 1532–1541.

24 Holmes, D.G., Lipo, T.A., McGrath, B.P., and Kong, W.Y. (2009). Optimized design of stationary frame three phase AC current regulators. *IEEE Trans. Power Electron.* 24 (11): 2417–2425.

25 Dannehl, J., Liserre, M., and Fuchs, F.W. (2011). Filter-based active damping of voltage source converters with LCL filter. *IEEE Trans. Ind. Electron.* 58 (8): 3623–3633.

26 McGrath, B.P., Parker, S.G., and Holmes, D.G. (2013). High performance current regulation for low-pulse-ratio inverters. *IEEE Trans. Ind. Appl.* 49 (1): 149–158.
27 Buso, S. and Mattavelli, P. (2006). *Digital Control in Power Electronics*. San Francisco, CA: Morgan & Claypool Publishers.
28 Yepes, A.G., Freijedo, F.D., Doval-Gandoy, J. et al. (2010). Effects of discretization methods on the performance of resonant controllers. *IEEE Trans. Power Electron.* 25 (7): 1692–1712.

10

Enhancement of System Stability Using Active Cancelation to Eliminate the Effect of Grid Impedance on System Stability and Injected Power Quality of Grid-Connected Inverter

10.1 Introduction

As described in Chapter 9, the introduction of *LC*-trap voltage or current feedback to reshape the inverter output impedance can effectively tackle stability issue. However, the circuit complexity of the entire system is increased. Distributed power generation (DPG) inverters installed at the same point of common coupling (PCC) are coupled to different devices. The grid impedance, which causes a voltage difference from the mains voltage, can cause harmonic interaction and current circulation between them [1–3].

This chapter proposes a grid impedance cancelator that aims to eliminate multiple resonance phenomena caused by the grid impedance. The concept is based on using a bidirectional dc–ac converter connected in series with the inverter output to create a negative virtual grid impedance to counteract the effect of the grid impedance.

By considering the Nyquist stability criteria, the operating principle of the grid impedance cancelator is firstly introduced. Several issues related to the realization of the cancelator will be given. The output voltage of the cancelator eliminates the harmonic voltage on the grid voltage, which is caused by the grid impedance. To improve the controllability and stability of the system, including dc link voltage control, ac voltage regulation, and active damping function, various control strategies are discussed. A model that describes the system operation with the effect of digital delay included will be given. The performance of a low-voltage prototype will be given. Results reveal that the cancelator does not only enhance harmonic rejection capability of the inverter against distorted grid voltage, but also improve the power quality of the injected current. More importantly, the volt-ampere (VA) rating of the cancelator is much smaller than that of the entire system. It constitutes a small amount of power consumption.

Control and Filter Design of Single-Phase Grid-Connected Converters, First Edition. Weimin Wu, Frede Blaabjerg, Henry Chung, Yuanbin He, and Min Huang.

Published 2023 by John Wiley & Sons, Inc.

10.2 Principle of the Grid Impedance Cancelator

The operating principle of the grid impedance cancaletor is given in this section. The system stability is studied by applying impedance-based stability analysis, which has been widely used to evaluate the small signal stability of DPG inverter systems. Figure 10.1 shows the equivalent circuit of a single-phase inverter in a DPG system. The inverter consists of a voltage source inverter (VSI), which generates pulsewidth modulated voltage pulses u_i, and an LCL filter formed by L_1, L_2, and C_f. It is connected to the grid voltage u_g with the grid impedance of Z_g. Its output current i_o is controlled.

Based on the impedance-based stability criterion, the inverter is represented by a small-signal current source ΔI_s in parallel with the output impedance Z_o, as shown in Figure 10.2. Assuming that both u_g and i_s are stable, i_o is expressed as [4].

$$\Delta I_o(s) = \left[\Delta I_s(s) - \frac{\Delta U_g(s)}{Z_o(s)}\right] \cdot \frac{1}{1 + Z_g(s)/Z_o(s)} \tag{10.1}$$

Based on (10.1), the effect of the grid impedance can be eliminated by either making the equivalent output impedance of the inverter (Z_o^*) infinite, as

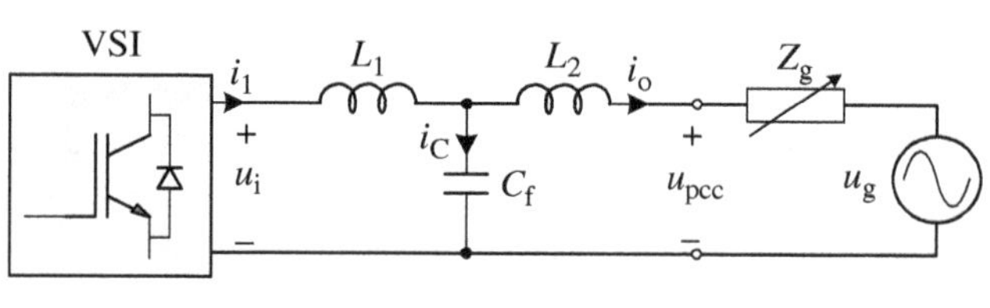

Figure 10.1 Equivalent circuit of a single-phase inverter in a DPG system.

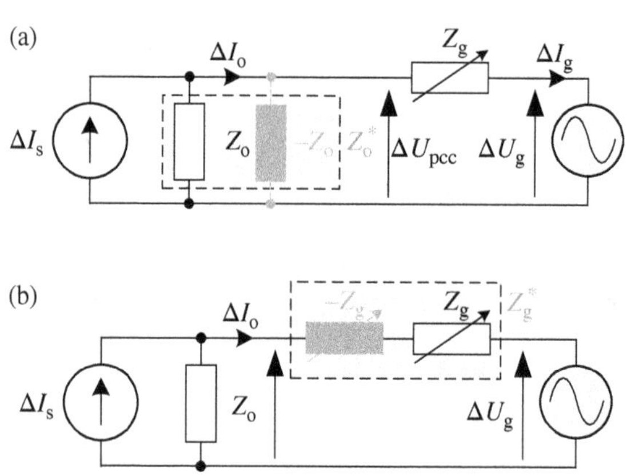

Figure 10.2 Illustration of the interactive decoupling between the inverter and the grid. (a) Infinite increase of the inverter output impedance. (b) Cancellation of equivalent grid impedance.

shown in Figure 10.2a, or the equivalent grid impedance (Z_g^*) zero, as shown in Figure 10.2b. The Nyquist stability criterion will always be satisfied and the system will always be stable regardless of the grid impedance. With Z_o^* tends to infinite, the reshaped grid current is:

$$\Delta i_g(s)\big|_{Z_o^*(s)\to\infty} = \left[\Delta I_s(s) - \frac{\Delta U_g(s)}{Z_o^*(s)}\right]\cdot\frac{1}{1+Z_g(s)/Z_o^*(s)} = \Delta I_s(s) \tag{10.2}$$

With Z_g^* tends to zero, the reshaped injected current i_o is:

$$\Delta i_o(s)\big|_{Z_g^*(s)\to 0} = \left[\Delta I_s(s) - \frac{\Delta U_g(s)}{Z_o(s)}\right]\cdot\frac{1}{1+Z_g^*(s)/Z_o(s)} = \Delta I_s(s) - \frac{\Delta U_g(s)}{Z_o(s)} \tag{10.3}$$

The effect of the grid impedance is then decoupled.

Figures. 10.3 a, b show the control block diagrams that realize (10.2) and (10.3) in a single-phase grid-connected inverter with single-loop direct control strategy. The output filter is of LCL type. The expression of the output impedance of the

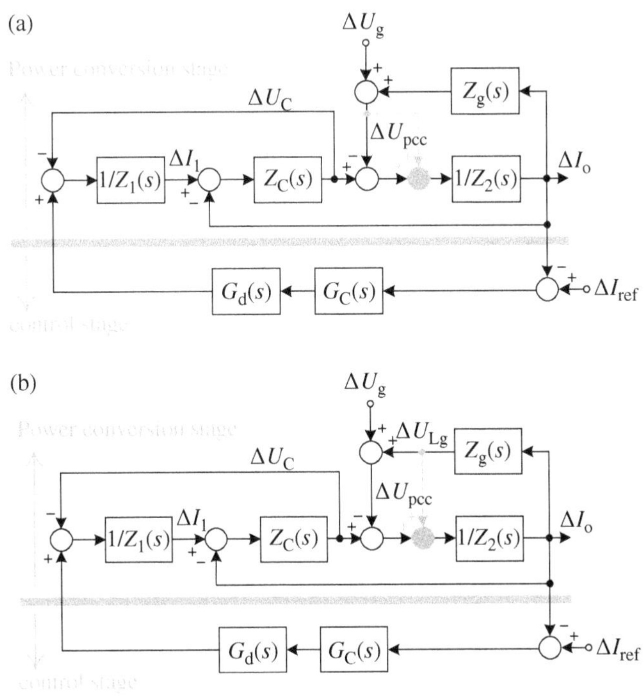

Figure 10.3 Block diagram of the decoupled *LCL* filter-based grid-connected inverter. (a) Infinite increase of the inverter output impedance. (b) Cancellation of equivalent grid impedance.

inverter Z_o is given in (10.4). The current sources, given in (10.2) and (10.3), are generalized in the form given in (10.5). Z_1, Z_2, and Z_C are the impedances of the inverter-side inductor, grid-side inductor, and filter capacitor of the *LCL* filter, respectively. G_C and G_d are the digital current controller and the digital delay, respectively. ΔI_{ref} is the current reference.

$$Z_o(s)\big|_{\Delta I_s(s)=0, Z_s(s)=0} = \left.\frac{\Delta U_g(s)}{\Delta I_o(s)}\right|_{\Delta I_s(s)=0, Z_q(s)=0} = \frac{Z_1(s)Z_2(s)+Z_1(s)Z_C(s)+Z_2(s)Z_C(s)+Z_C(s)G_C(s)G_d(s)}{Z_1(s)+Z_C(s)} \tag{10.4}$$

$$\Delta I_S(s) = \frac{Z_C(s)G_C(s)G_d(s)}{Z_1(s)Z_2(s)+Z_1(s)Z_C(s)+Z_2(s)Z_C(s)+Z_C(s)G_C(s)G_d(s)} \cdot \Delta I_{ref}(s) \tag{10.5}$$

The methods of inserting a fictious impedance depicted in Figure 10.3 are realized by a voltage-controlled voltage source connected in series with the output of the inverter. The realization is given in Figure 10.4, in which the notch filter with

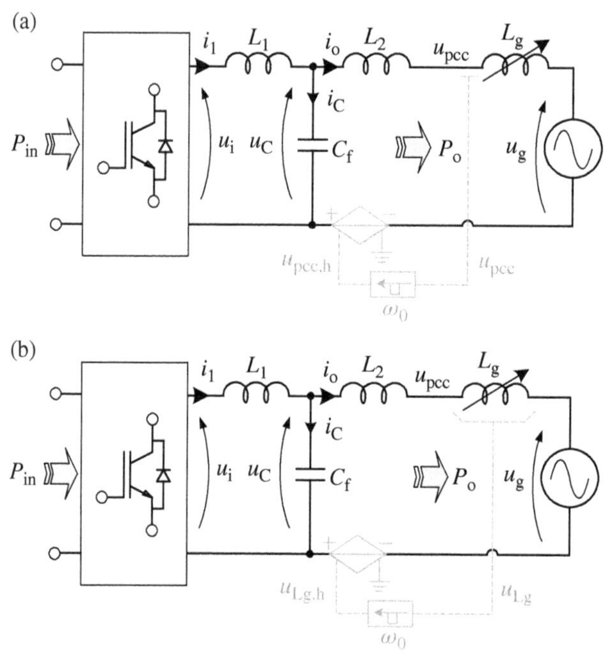

Figure 10.4 Illustration of the grid decoupling mechanism. (a) Infinite increase of the inverter output impedance. (b) Cancellation of equivalent grid impedance.

the center frequency ω_o is used to eliminate the dominant line-frequency (ω_o) component. $u_{pcc,\,h}$ and $u_{L_{g,h}}$ represent the harmonic voltage at the PCC and across the equivalent grid impedance, respectively. In order to study the worst-case scenario and simplify the analysis, the resistive part in the power filter and the grid impedance are neglected in Figure 10.4 and in the following discussions. As u_{L_g} is not measurable, cancelation of the equivalent grid impedance, as shown in Figure 10.4b, cannot be achieved directly.

The difference between Figures 10.4a, b is that the output voltage of the cancelator is reference from different voltages u_{pcc} and u_{L_g}. Thus, it can be seen that

$$\Delta U_{pcc}(s) = \Delta U_{L_g}(s) + \Delta U_g(s) \tag{10.6}$$

It should be noted that the grid voltage source ΔU_g has no effect on the stability. Thus, increasing the output impedance of the inverter is similar to canceling the grid impedance and harmonic voltage on u_g. Hence, both approaches can be achieved by using the same circuit in Figure 10.4a. As the implementation is simply based on feedforwarding the PCC voltage to realize grid decoupling, it is independent on the circuit parameters of the inverter.

10.3 Modeling with the Grid Impedance Cancelator

10.3.1 System Configuration with the Grid Impedance Cancelator

Figure 10.5 shows the connection of the cancelator to a grid-connected photovoltaic (PV) inverter with LCL filter. The grid voltage is 220 Vac, 50 Hz. The dc voltage of the inverter is 400 V. The cancelator consists of a full-bridge dc–ac converter without output filter. Its dc side consists of a dc capacitor with the voltage ranging from 30 to 50 V, thereby low-voltage switching devices can be used. Its average value of the output voltage u_{ic} is equal to the harmonic voltage at the PCC, $u_{pcc,\,h}$, as described in Section 10.2.

Several issues related to the implementation of the cancelator are addressed as follows:

1) The voltage rating of the switching devices required is much lower than that used in the inverter. Thus, devices of low on-state resistance can be used, implying that the conduction and switching losses can be reduced [5].
2) Figure 10.6 shows the key waveforms of the entire system. As the switching frequency of the cancelator is much higher than that of the inverter, the ripple current on the grid current is very small. Thus, the core loss of the grid-side filtering inductor L_2 is very small.

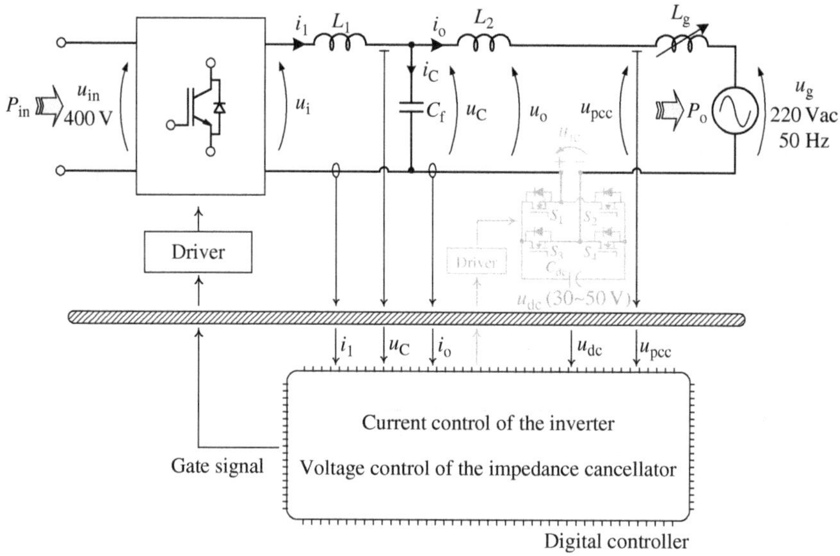

Figure 10.5 Proposed single-phase inverter system configuration with the grid impedance cancelator.

Figure 10.7a shows the equivalent circuit of the proposed inverter system. The output voltage of the cancelator u_{ic} is equal to the harmonic voltage at the PCC $u_{pcc,\,h}$ at the steady state. Thus, as shown in Figure 10.7b, it consists of the harmonic voltage of the grid voltage $u_{g,\,h}$ and that across the equivalent grid impedance $u_{L_g,h}$. The injected harmonic voltage $-u_{L_g,h}$ is used to counteract the voltage drop across the equivalent grid impedance and suppress harmonic disturbances generated by other local DPG systems at the PCC.

10.3.2 AC Voltage Regulation

u_{ic} is the voltage from the full bridge without an output filter. As shown in Figure 10.8, the harmonic voltage at the PCC is sampled, downscaled, and feed-forwarded into the modulator to generate the necessary voltage. The notch filter with the center frequency of ω_o is used to eliminate the dominant line frequency component of the PCC voltage. T_S is the sampling period which is synchronized with the switching period.

10.3.3 Active Damping Function

Active damping function has been widely applied to the current control loop in the inverter for suppressing filter resonance. However, its function is affected by the

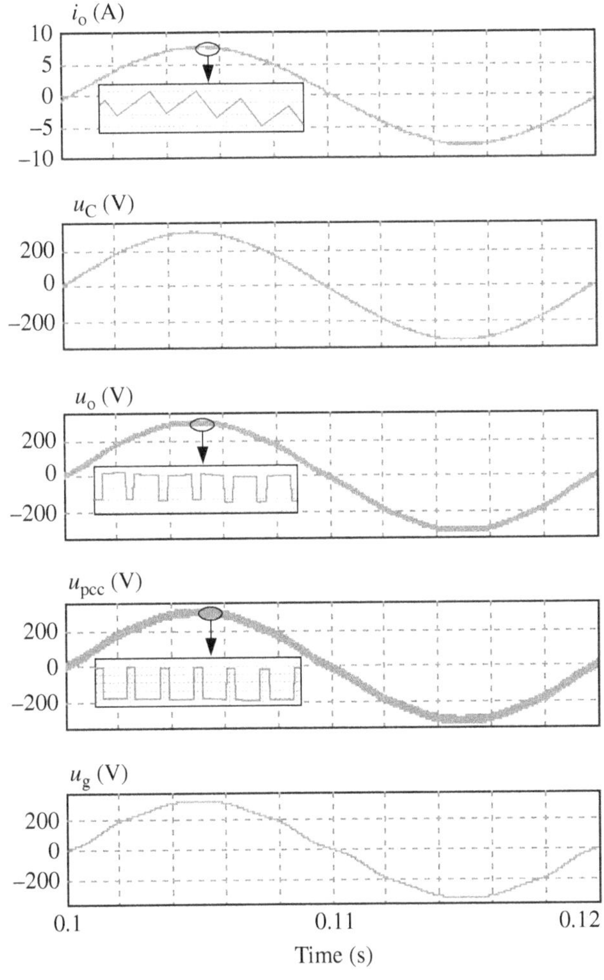

Figure 10.6 Key waveforms.

sample-to-update time delay. Since the digital delay of the cancelator is significantly reduced because the sampling rate is much higher than the inverter. Active damping function can be integrated into the cancelator, so as to widen the effective range and lessen the computational burden of the inverter controller. It is implemented using a simple proportional harmonic injected current feedback. It is shown in Figure 10.8. R_v is the gain of the active damping function or can be interpreted as the virtual resistance.

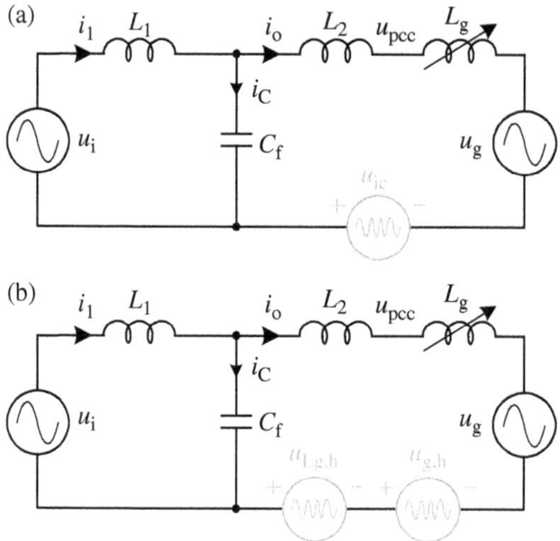

Figure 10.7 Simplified equivalent circuit of the system. (a) Simplified equivalent voltage source u_{ic}. (b) Simplified equivalent split voltage sources $u_{Lg,h}$ and $u_{g,h}$.

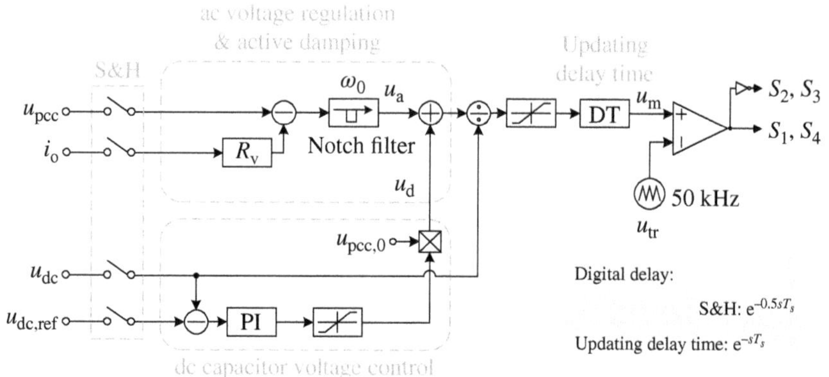

Figure 10.8 Control mechanism of the cancelator.

The expected value of R_v is:

$$R_v = \frac{\omega_{r\_up} L_2 + \omega_{r\_down}(L_1 + L_2)}{2} \tag{10.7}$$

where $\omega_{r\_down} = \sqrt{\frac{1}{L_1 C_f}}$ and $\omega_{r\_up} = \sqrt{\frac{L_1 + L_2}{L_1 L_2 C_f}}$

Equation (10.7) is derived as follows.

Based on Figure 10.12, the characteristic resonant frequency of the filter is changeable with different values of R_v. For example, when $R_v = 0$,

$$\omega_{r\_up} = \sqrt{\frac{L_1 + L_2}{L_1 L_2 C_f}} \tag{10.8}$$

where ω_{r_up} is the maximum characteristic frequency.

When $R_v \to \infty$,

$$\omega_{r\_down} = \sqrt{\frac{1}{L_1 C_f}} \tag{10.9}$$

where ω_{r_down} is the minimum characteristic frequency.

The impedance of L_1 and branch $L_2 - R_v$, is calculated as:

$$Z_{L(LCL)} = sL_1 \frac{sL_2 + R_v}{s(L_1 + L_2) + R_v} \tag{10.10}$$

with the impedance characteristic shown in Figure 10.9.

To determine the optimized quality factor (*Q*-factor), the range of the changeable characteristic frequency between (10.8) and (10.9) should fall into the shadow areas in Figure 10.9. It is the resistive region of L_1 and $L_2 - R_v$ branch [6]. Based on (10.10), the virtual resistance is confined within the range of:

$$\frac{R_v}{L_1 + L_2} \le \omega_r \le \frac{R_v}{L_2} \tag{10.11}$$

Let the virtual resistance R_v change from 0 to ∞ Ω. It can be deduced that:

$$\omega_{r\_down} \le \omega_r \le \omega_{r\_up} \tag{10.12}$$

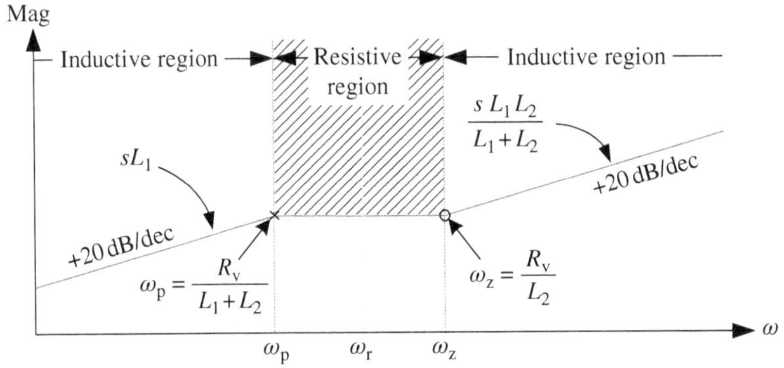

Figure 10.9 Impedance feature of L_1 and branch $L_2 - R_v$.

Combining (10.11) and (10.12), the selection criterion of the apparent virtual resistance is:

$$\begin{cases} \dfrac{R_r}{L_1 + L_2} \leq \omega_{r\_down} \\ \dfrac{R_v}{L_2} \geq \omega_{r\_up} \end{cases} \tag{10.13}$$

The optimized value of R_v is given as (10.7).

10.3.4 DC Capacitor Voltage Control

The dc capacitor voltage is regulated at the reference $u_{dc,ref}$ by a closed-loop feedback control [7]. $u_{dc,ref}$ is taken to be 30 V in the prototype. As shown in Figure 10.8, the dc capacitor voltage is sensed, downscaled, and subtracted by $u_{dc,ref}$. The error is fed to a proportional-plus-integral (PI) amplifier with its output multiplied by a sinusoidal signal, which is obtained by a phase-locked loop with the grid voltage as reference. The generated output is added to the modulating information. In other words, the dc capacitor voltage is regulated by changing the real power flowing into the capacitor.

The cancelator absorbs high-frequency components to emulate the resistive component R_v and counteracts the effects of the grid impedance on the inverter. The absorbed high-frequency power is then transformed to low-frequency one by forming a voltage in phase with the line voltage, as illustrated in Figure 10.8, to maintain u_{dc} at $u_{dc,ref}$. The absorbed real power is to compensate the power dissipation of the entire system and, at the same time, emulate the resistive component R_v.

10.4 Modeling of the Grid Impedance Cancelator

By combining the ac voltage regulation and the active damping function, as shown in Figure 10.8, the output u_a is:

$$\Delta U_a(s) = G_{NF}(s)e^{-0.5sT_S}\left[\Delta U_{pcc}(s) - R_v \Delta I_o(s)\right] \tag{10.14}$$

where $e^{-0.5sT_S}$ is the transfer function of the sampling-and-holding unit, and $G_{NF}(s)$ represents the notch filter.

$G_{NF}(s)$ is expressed as:

$$G_{NF}(s) = \frac{s^2 + \omega_o^2}{s^2 + 2\omega_b s + \omega_o^2} \tag{10.15}$$

The center line frequency ω_o is generated by the PLL. The response time of the filter is determined by the band ω_b.

The output signal u_d in the dc capacitor voltage feedback loop is:

$$\Delta U_d(s) = \Delta U_{pcc,0}(s) G_{PI}(s) e^{-0.5sT_s} \times [\Delta U_{dc}(s) - \Delta U_{dc,ref}(s)] \tag{10.16}$$

where $G_{PI}(s)$ is the transfer function of PI controller and the sinusoidal signal $\Delta U_{pcc,\,0}$ is used to force the dominant line-frequency component at ΔU_d.

The cutoff frequency of the PI controller is designed to be much lower than the line frequency. By combining (10.14) and (10.16), the modulating signal ΔU_m is expressed as:

$$\Delta U_m(s) = \frac{\Delta U_a(s) + \Delta U_d(s)}{U_{dc}} e^{-sT_s} \tag{10.17}$$

The output voltage of the impedance cancelator is:

$$\Delta U_{ic}(s) = [\Delta U_a(s) + \Delta U_d(s)] e^{-sT_s} \tag{10.18}$$

Consider that the control bandwidth of the dc capacitor voltage feedback loop is designed to be much lower than the line frequency. ΔU_d can be neglected. Equation (10.18) can be simplified into:

$$\begin{aligned} \Delta U_{ic}(s)|_{\Delta U_d(s) \cong 0} &= \Delta U_a(s) e^{-sT_s} \\ &= G_{NF}(s) e^{-1.5sT_s} \left[\Delta U_{pcc}(s) - R_v \Delta I_o(s)\right] \\ &= e^{-1.5sT_s} \left[\Delta U_{pcc,h}(s) - R_v \Delta I_{o,h}(s)\right] \end{aligned} \tag{10.19}$$

Based on (10.19), the equivalent circuit is obtained by separating the ac voltage regulation (or the equivalent grid impedance cancelation) and the active damping function. This is shown in Figure 10.10a. The corresponding physical meaning can be represented by the equivalent impedance, as shown in Figure 10.10b. $u_{g,\,0}$ represents the line-frequency component of the grid voltage. The virtual resistance, $R_{v,h} = R_v e^{-1.5sT_s}$, is included by the active damping function. The negative grid impedance, $L_{g,h} = L_g e^{-1.5sT_s}$, is formed by the unity gain-based PCC harmonic voltage feedforward.

Both virtual resistance and negative grid impedance have the digital delay term of $e^{-1.5sT_s}$. To examine the effect of the digital delay on them, the mathematical model of the original *LCL* filter and the reshaped one with respect to the grid impedance are derived and compared. Based on Figure 10.1, the open-loop transfer function of the original *LCL* filter with respect to the grid impedance is expressed as:

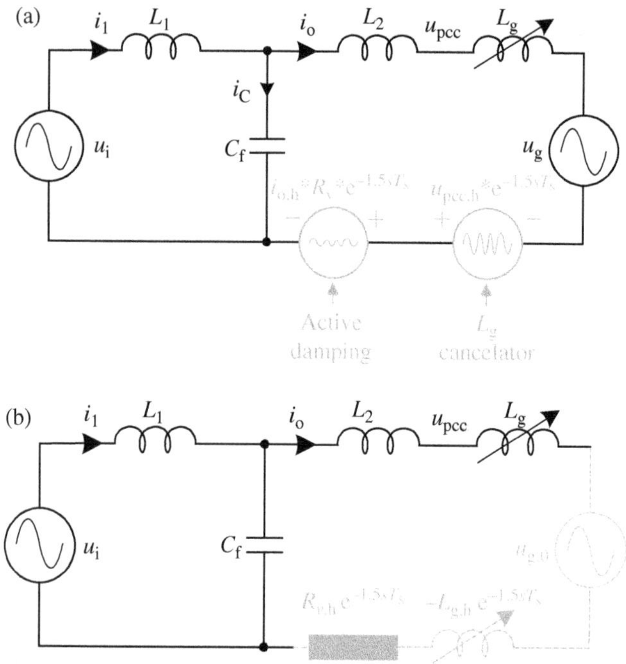

Figure 10.10 Equivalent circuit of the grid impedance cancelator. (a) Equivalent voltage source. (b) Equivalent impedance.

$$G_{o,o}(s) = \frac{\Delta I_o(s)}{\Delta U_i(s)}$$
$$= \frac{Z_C(s)}{Z_1(s)\left[Z_2(s) + Z_g(s)\right] + Z_1(s)Z_C(s) + \left[Z_2(s) + Z_g(s)\right]Z_C(s)} \tag{10.20}$$

After reshaping the impedance, the inserted series impedance, due to the grid impedance cancelator, is:

$$Z_{ic}(s) = G_{NF}(s)\left(R_v e^{-1.5sT_s} - L_g e^{-1.5sT_s}\right) \tag{10.21}$$

The open-loop transfer function of the reshaped "*LCL* filter" is expressed as:

$$G_{o,ic}(s) = \frac{\Delta I_o(s)}{\Delta U_i(s)}$$
$$= \frac{Z_C(s)}{Z_1(s)\left[Z_2(s) + Z_{ic}(s) + Z_g(s)\right] + Z_1(s)Z_C(s) + \left[Z_2(s) + Z_{ic}(s) + Z_g(s)\right]Z_C(s)} \tag{10.22}$$

By replacing the filter parameters with the value tabulated in Table 10.1, Eqs. (10.20) and (10.22) are given in Figure 10.11, in which a set of frequency

Table 10.1 Values of the components used in the inverter prototype.

	LCL filter			Grid inductance	DC capacitor	Switching frequency		DC input voltage	DC capacitor voltage	PCC voltage
System parameter	L_1	L_2	C_f	L_g	C_{dc}	f_{sw-1} (Inverter)	f_{sw-2} (Cancelator)	u_{in}	u_{dc}	u_{pcc}
Value	2 mH	1 mH	4.3 μF	0 ~ 7.2 mH	4 × 680 μF	20 kHz	50 kHz	220 V	30 V	125 V

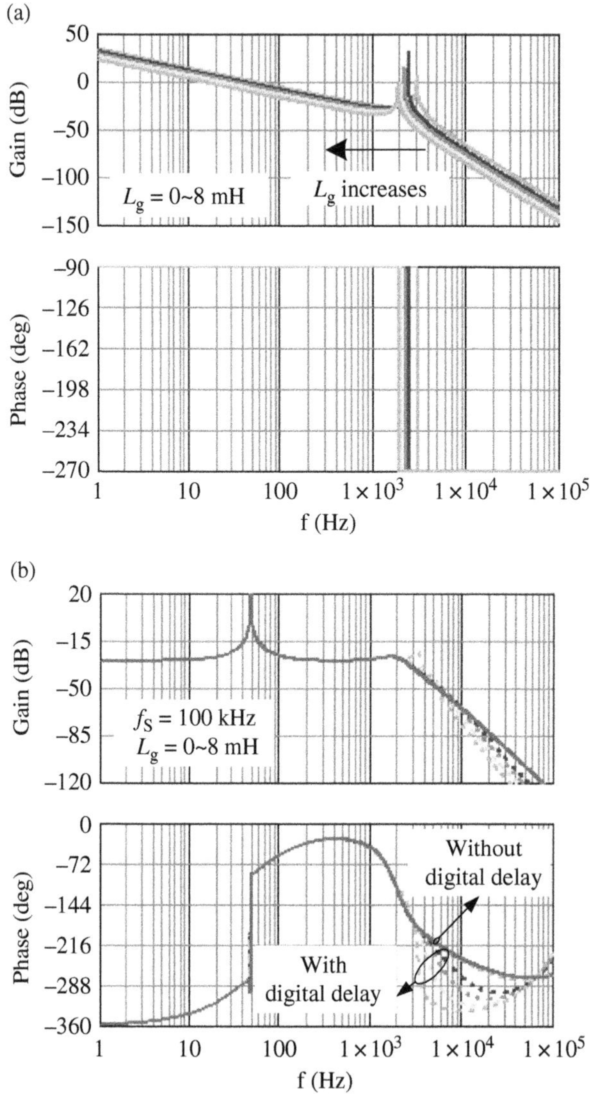

Figure 10.11 Open-loop transfer functions of the original *LCL* filter and the reshaped one with variable grid impedance. (a) Original *LCL* filter. (b) Replaced "*LCL* filter".

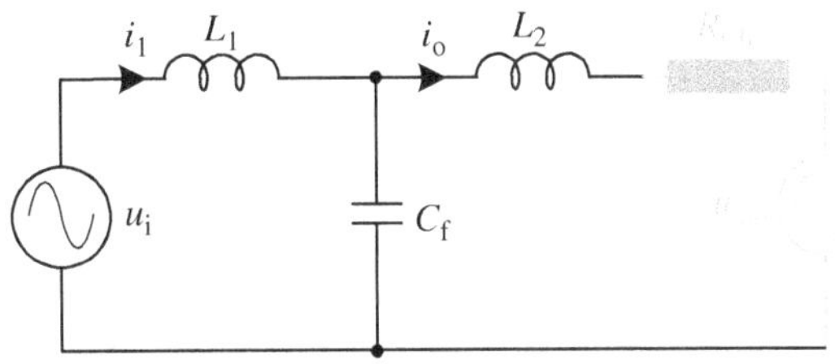

Figure 10.12 Reshaped well-damped grid-connected inverter with the cancelation of L_g.

response of the reshaped "*LCL* filter" with the grid impedance cancelation is overlaid with each other. The following points can be noted:

1) without the grid impedance cancelator, the filter resonant frequency varies widely with the equivalent grid impedance;
2) with the grid impedance cancelator included, the effect of equivalent grid impedance is eliminated;
3) with the active damping function, the filter resonance is well damped; and
4) the digital delay leads to minor impact on the impedance reshaping within the filter resonant frequency, as shown in Figure 10.11b.

Therefore, by neglecting the effect of minor digital delay, the reshaped grid-connected inverter with an *LCL* filter behaves as a well-damped one with the equivalent grid impedance canceled, as shown in Figure 10.12.

10.5 Experimental Verification

An experimental setup consisting of a single-phase inverter connected in series with a grid impedance cancelator to the power grid is evaluated. The design parameters are tabulated in Table 10.1. The controller of the inverter is Texas Instruments TMS320F2808. The inverter is supplied by an external power supply KIKUSUI PCR3000LE and the output voltage of the inverter is 125 V, 50 Hz. It is connected to the grid of 220 V, 50 Hz via an isolator. The experimental inverter setup with variable grid inductance and the impedance cancelator is presented in Figure 10.13. The leakage inductance is 0.8 mH. The emulated grid inductance is variable.

The cancelator is designed by considering the volt-ampere (VA) rating. The current rating of the switching devices and the dc capacitors is chosen by considering the inverter output current. The voltage rating is determined by the maximum harmonic PCC voltage under the worst-case condition (e.g., the voltage THD is less than 5%). To avoid square pulse voltage superimposed onto the PCC voltage, as shown in Figure 10.6, an optional $R_d - C_d$ branch is connected at the PCC.

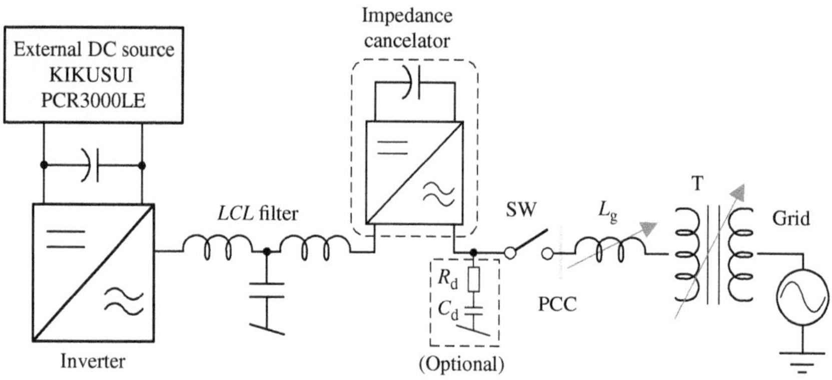

Figure 10.13 Experimental inverter setup with variable grid inductance.

The inverter is regulated by proportional-resonant-plus-harmonic-compensation-based current control strategy with unity gain-controlled filter capacitor voltage feedforward [8].

Figure 10.14a shows the key waveforms when the grid-connected inverter is disabled and the cancelator is disabled. Due to the standby *LC* resonant tank at the grid side of the inverter, the oscillation appears at the grid current and the PCC voltage. Figure 10.14b shows the corresponding waveforms when the cancelator is enabled. The oscillation is not present, because of the inserted active damping function or virtual resistance. It also prevents the voltage at the inverter output from harmonics.

Figures 10.15–10.17 show the steady-state waveforms of the inverter outputs with and without the grid impedance cancelator under different grid impedances. It can be seen that the grid impedance cancelator enhances the harmonic rejection capability of the inverter against the grid voltage distortion and improves the power quality of the injected current a lot.

Without the grid impedance cancelator, the inverter becomes unstable as the equivalent grid inductance L_g increases from 0 to 7.2 mH. They are shown in Figures 10.15a–10.17a. The measured total-harmonic-distortion (THD) of the injected current increases from 1.7, 2.7, to 6.6%. Figure 10.18 presents the corresponding frequency spectrum of the oscillating injected current when $L_g = 7.2$ mH, where the oscillation around 0.7 kHz is caused by the poor stability margin. The result matches with the theoretical stability analysis in [8].

With the grid impedance cancelator engaged, the inverter is maintained stable even under wide variation of equivalent grid impedance, as shown in Figures 10.15b–10.17b, because of elimination of equivalent grid impedance and

(a)

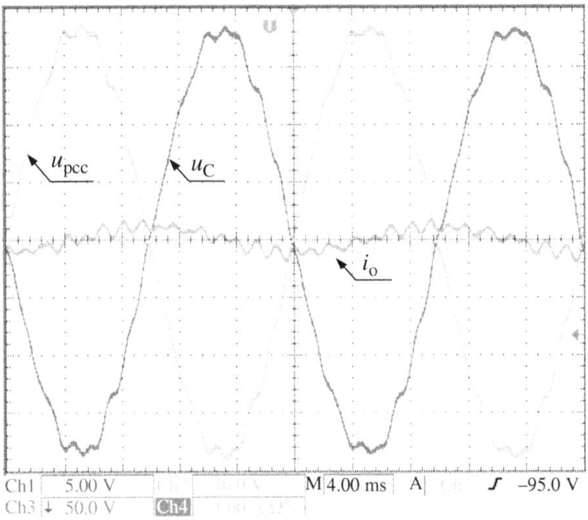

(b)

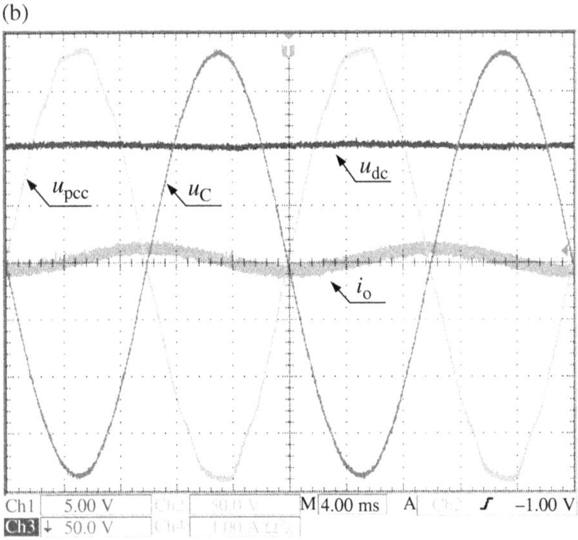

Figure 10.14 Key waveforms with the inverter disabled. (a) Cancelator disabled. (b) Cancelator enabled.

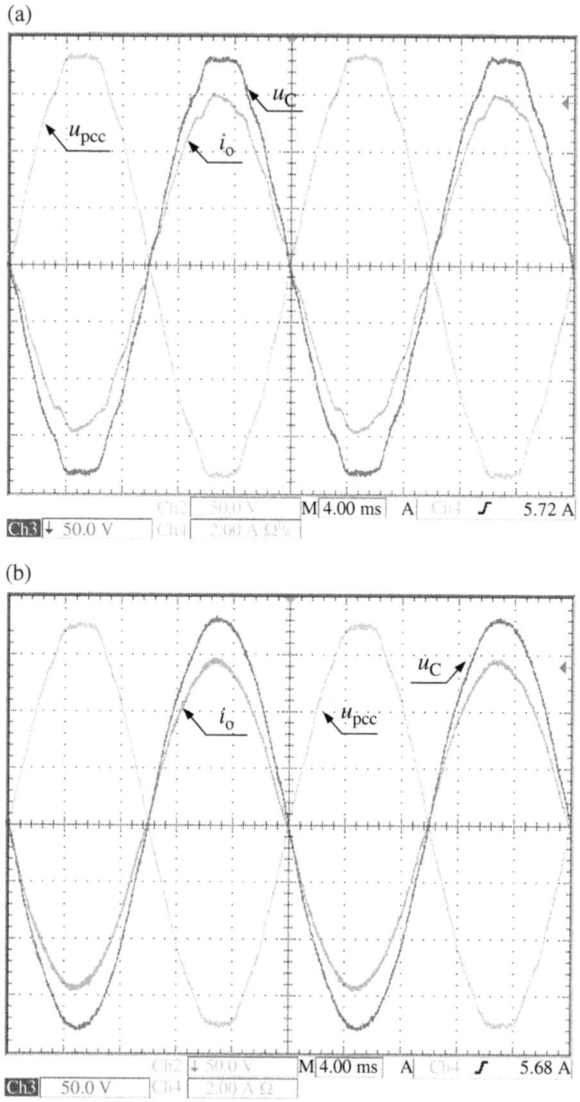

Figure 10.15 Steady-state waveforms of the inverter outputs when $L_g = 0$. (a) Cancelator disabled. (b) Cancelator enabled.

(a)

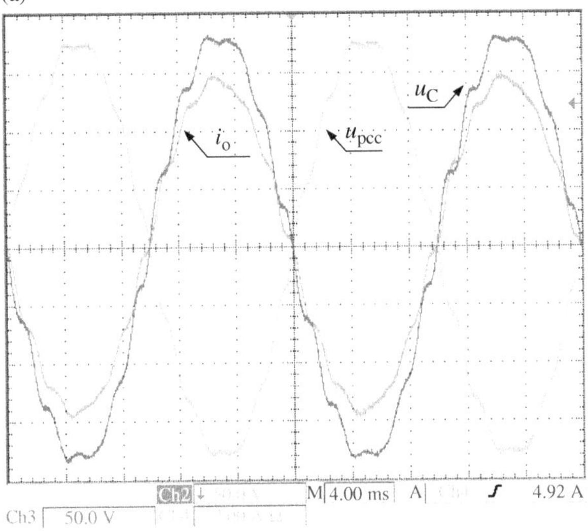

(b)

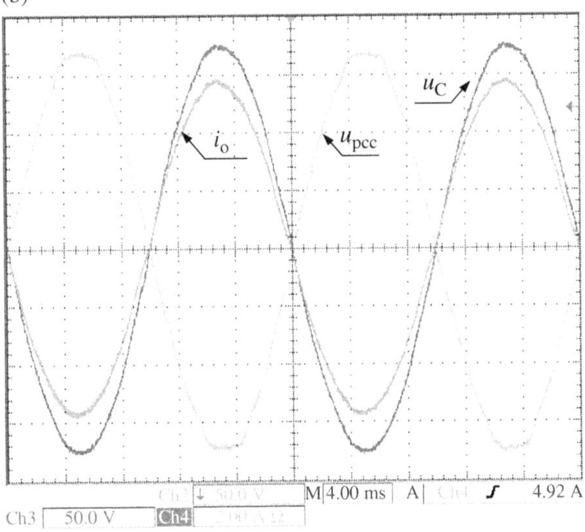

Figure 10.16 Steady-state waveforms of the inverter outputs when L_g = 2 mH. (a) Cancelator disabled. (b) Cancelator enabled.

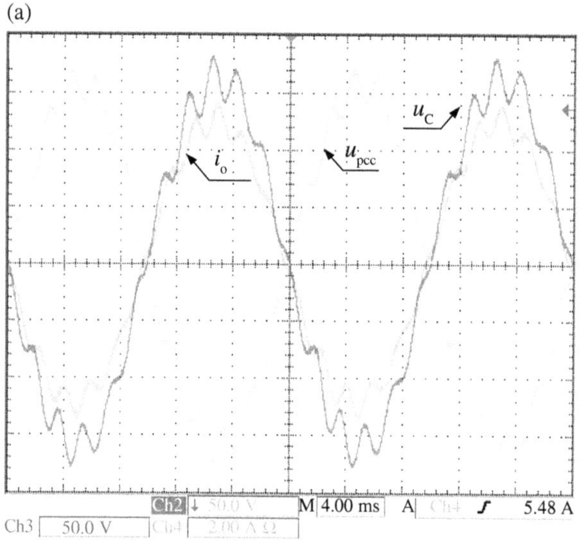

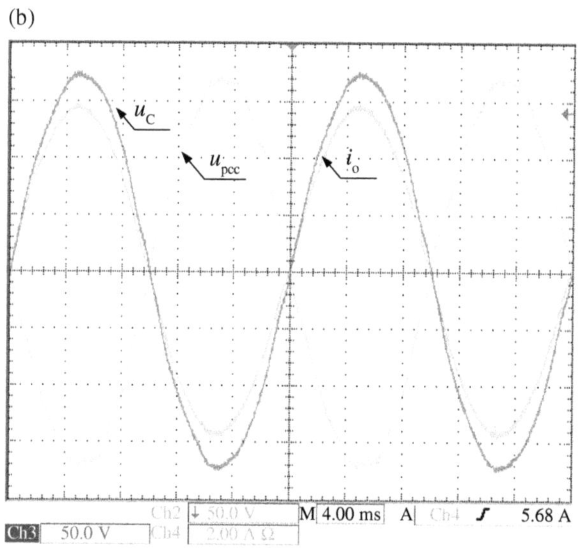

Figure 10.17 Steady-state waveforms of the inverter outputs when L_g = 7.2 mH. (a) Cancelator disabled. (b) Cancelator enabled.

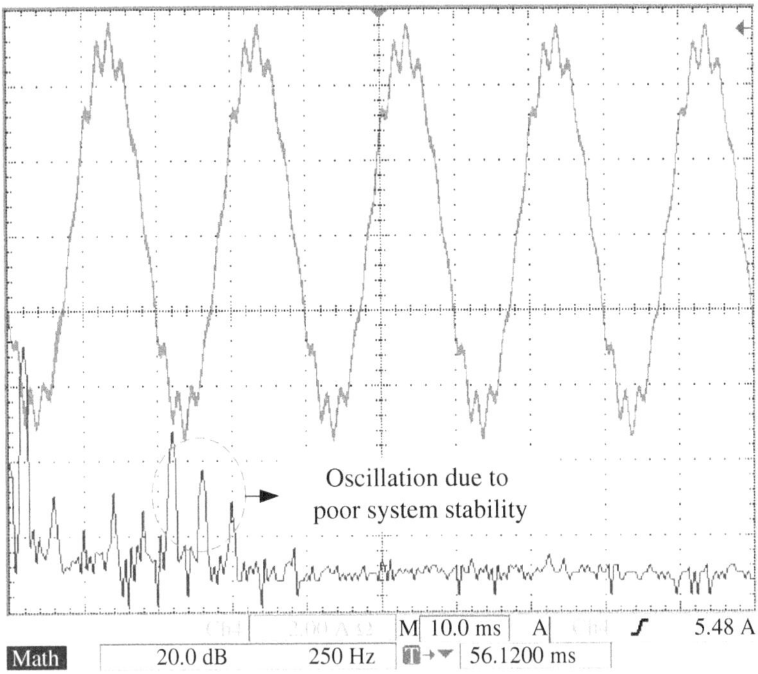

Figure 10.18 Frequency spectrum of the injected current.

the insertion of virtual resistance at the grid side, as shown in Figure 10.12. The measured THD of the injected current is at 0.9%, 1.0%, and 0.8%, respectively.

When the cancelator is engaged and the power rating is suddenly changed from half to full power, and vice versa, the dynamic behavior of the cancelator is investigated under different grid inductances. The corresponding waveforms are shown in Figure 10.19. Figure 10.19a shows the waveforms when the cancelator is engaged. The system becomes stable with the cancelator engaged. Figure 10.19b also indicates that the inverter operates well with the power transients. The amplitude variation of the filter capacitor voltage during transient state is deal to the notch filter, as shown in Figure 10.8.

Considering the extreme low dc capacitor voltage, switching devices of the impedance cancelator are MOSFET FDD86110 (100 V, 50 A) with on-state resistance of 10.2 mΩ. Based on that, Figure 10.20 shows the power consumption of the cancelator (P_d) against the inverter input power (P_{in}).The power loss of the cancelator is mainly composed of the switching devices and dc capacitors. It accounts for lower than 0.5% of the inverter input power.

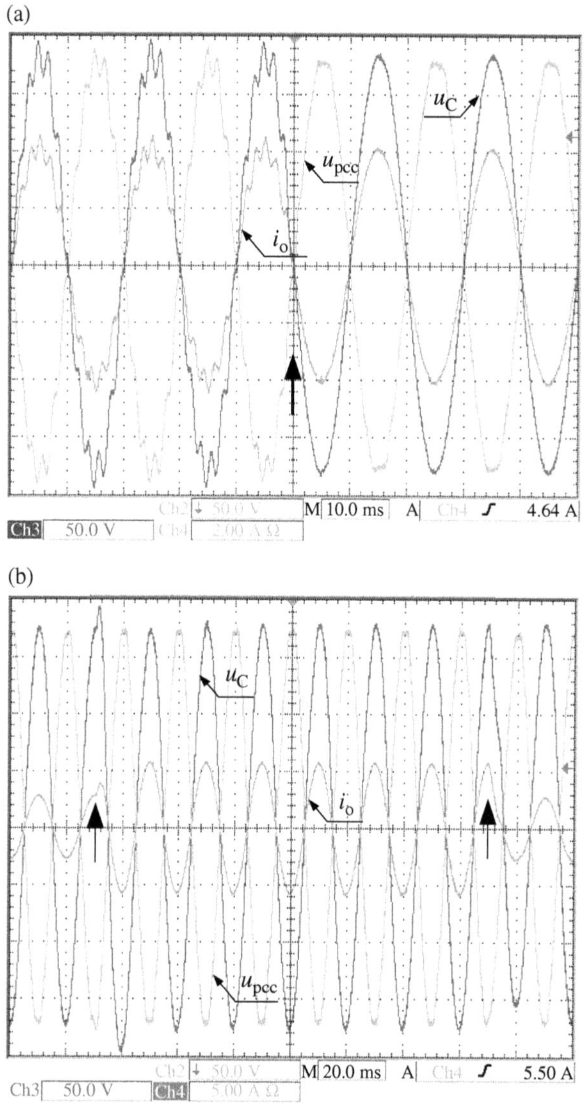

Figure 10.19 Dynamic behaviors of the grid-connected inverter when L_g = 7.2 mH. (a) Cancelator from disabled to enabled. (b) Injected power transients.

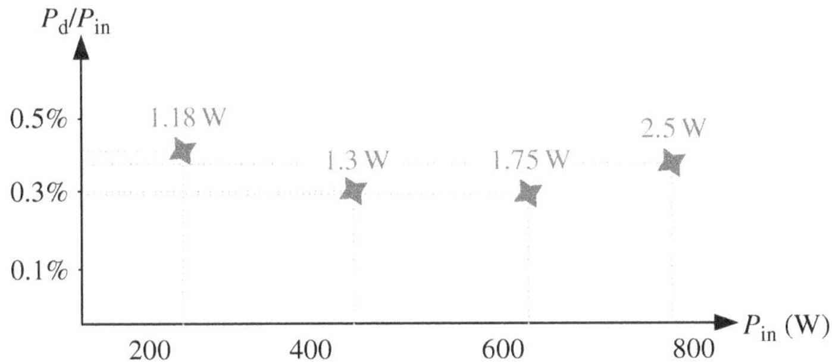

Figure 10.20 Power loss of the cancelator as a function of the inverter input power.

10.6 Summary

In this chapter, cancelation of equivalent grid impedance with a full-bridge dc–ac converter has been discussed. Its basic principle is derived by considering the impedance-based stability criterion. It has been proven that the method exhibits similar characteristics with the increase in the inverter output impedance by introducing a negative virtual grid impedance. The grid impedance cancelator features the following:

1) unnecessary to use extra passive *LC* filter elements;
2) cancelation of equivalent grid impedance;
3) strong harmonic suppression against grid disturbance; and
4) active damping function included to reduce the computational complexity of the inverter controller.

A single-phase testbed, consisting of a grid-connected inverter and variable grid inductance, has been built and evaluated. Results show that the cancelator can improve the stability of the inverter and the power quality of the injected current even under wide variation of grid conditions. Compared with the traditional inverter, the reshaped inverter with the grid impedance cancelator performs as a well-damped one without equivalent grid impedance.

References

1 Agorreta, J.L., Borrega, M., Lopez, J., and Marroyo, L. (2011). Modeling and control of N-paralleled grid-connected inverters with LCL filter coupled due to grid impedance in PV plants. *IEEE Trans. Power Electron.* 26 (3): 770–785.

2 Enslin, J. and Heskes, P. (2004). Harmonic interaction between a large number of distributed power inverters and the distribution network. *IEEE Trans. Power Electron.* 19 (6): 1586–1593.

3 Wang, F., Duarte, J.L., Hendrix, M.A.M., and Ribeiro, P.F. (2011). Modeling and analysis of grid harmonic distortion impact of aggregated DG inverters. *IEEE Trans. Power Electron.* 26 (3): 786–797.

4 Sun, J. (2011). Impedance-based stability criterion for grid-connected inverters. *IEEE Trans. Power Electron.* 26 (11): 3075–3078.

5 Mohan, N., Undeland, T.M., and Robbins, W.P. (2003). *Power Electronics Converters, Applications, and Design*, 3e. Hoboken, NJ: Wiley.

6 Wu, W., He, Y., Tang, T., and Blaabjerg, F. (2013). A new design method for the passive damped *LCL*- and LLCL-filter based single-phase grid-tied inverter. *IEEE Trans. Ind. Electron* 60 (10): 4339–4350.

7 Pan, Z., Peng, F., and Wang, S. (2005). Power factor correction using a series active filter. *IEEE Trans. Power Electron.* 20 (1): 148–153.

8 He, Y., Wang, K., and Chung, H. (2014). Utilization of proportional filter-capacitor voltage feedforward to realize active damping for digitally-controlled gridtied inverter operating under wide grid impedance variation. Proceedings of the IEEE Energy Conversion Congress Exposition, pp. 4450–4457, Pittsburgh, PA (September 2014). New York: IEEE.

Index

Control and Filter Design of Single-Phase Grid-Connected Converters, First Edition. Weimin Wu, Frede Blaabjerg, Henry Chung, Yuanbin He, and Min Huang.

Published 2023 by John Wiley & Sons, Inc.

V

W

Z

www.ingramcontent.com/pod-product-compliance
Lightning Source LLC
LaVergne TN
LVHW020358191025
823407LV00001B/90

* 9 7 8 1 1 1 9 8 8 6 5 4 9 *